What's Happening
in the **Mathematical**
Sciences

Volume

What's Happening
in the Mathematical
Sciences

AMERICAN MATHEMATICAL SOCIETY
www.ams.org

2000 *Mathematics Subject Classification*:
00A06

ISBN-13: 978-0-8218-4478-6
ISBN-10: 0-8218-4478-4

For additional information and
updates on this book, visit
www.ams.org/bookpages/happening-7

The cover and the frontmatter for this publication were prepared using the Adobe®
CS2® suite of software. The articles were
prepared using TeX. TeX is a trademark of
the American Mathematical Society.

About the Author

DANA MACKENZIE is a freelance mathematics and
science writer who lives in Santa Cruz, California.
He received his Ph.D. in mathematics from Princeton
University in 1983 and taught at Duke University and
Kenyon College. In 1993 he won the George Pólya Award
for exposition from the Mathematical Association of
America. Changing his career path, he completed the
Science Communication Program at the University of
California at Santa Cruz in 1997. Since then he has
written for such magazines as *Science, New Scientist,
SIAM News, Discover,* and *Smithsonian,* and worked as
a contributing editor for *American Scientist.* His first
book, *The Big Splat, or How Our Moon Came to Be,* was
published by John Wiley & Sons and named to *Booklist*
magazine's Editor's Choice list for 2003. He co-wrote
Volume 6 of *What's Happening in the Mathematical
Sciences* with Barry Cipra.

Cover

A random tiling of a hexagon by lozenges. Each lozenge
is formed by two equilateral triangles glued together.
(Figure courtesy of Richard Kenyon.)

Contents

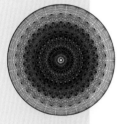

Introduction

WELCOME TO THE SEVENTH VOLUME of *What's Happening in the Mathematical Sciences.* Once again, it showcases the remarkable vitality of mathematics. Reading the articles in the book, you may be amazed by how much has happened in mathematics in the short time that has passed since the previous volume appeared.

One of the recurring features of the achievements described in the book is that major discoveries are often made where two or more well-established mathematical disciplines overlap. For example, in "A New Twist in Knot Theory," modular knots, which are already a mix of number theory and topology, are shown to be the same as knots that appear in dynamical systems and which are related to the chaotic behavior that makes precise weather prediction impossible. "Error-term Roulette and the Sato-Tate Conjecture" describes the delicate interplay between probability, which is inherently random, and number theory, which is traditionally deterministic and precise. "Dominos, Anyone?" connects the finitary discrete area of combinatorics with statistical mechanics and differential equations—disciplines that are infinitary and continuous. Mathematics and applications have had some extraordinary developments recently. "Not Seeing is Believing" shows how the dream of H.G. Wells depicted in *The Invisible Man* can come true and why the director of the next movie version should seek the help of mathematicians, as well as special effects artists. "The Fifty-one Percent Solution" describes the startling discovery that coin tossing is not a perfect model of randomness. Another surprising result, explained in "Compressed Sensing Makes Every Pixel Count," is that one can (and sometimes *should*) replace a 10 mega-pixel camera with a 30 *kilo*-pixel device, or even a one-pixel device.

Two articles touch upon further achievement in established mathematical areas. "Getting with the (Mori) Program" explains how the major obstacles to constructing the simplest (minimal) models of algebraic varieties of dimension three and higher have been removed. "Charting a 248-dimensional World" tells a fascinating story of how a large group of specialists in Lie theory, scattered all around the world, cooperated to uncover the buried structure of the Lie group E_8, one of the most complicated algebraic objects of its kind.

The techniques of CSI crime labs and particle physicists' linear accelerators come together to solve the mystery of "The Book that Time Couldn't Erase." This is the story of what is known as the Archimedes Palimpsest, a 13th century copy of the original text of Archimedes, which now opens a new link to the mind of the great mathematician of ancient Greece.

All in all, we hope that when you finish reading the book you share the excitement we find in the vibrant, active science of mathematics. And, most of all, we hope you enjoy reading it.

Sergei Gelfand, Publisher
Edward Dunne, Editor

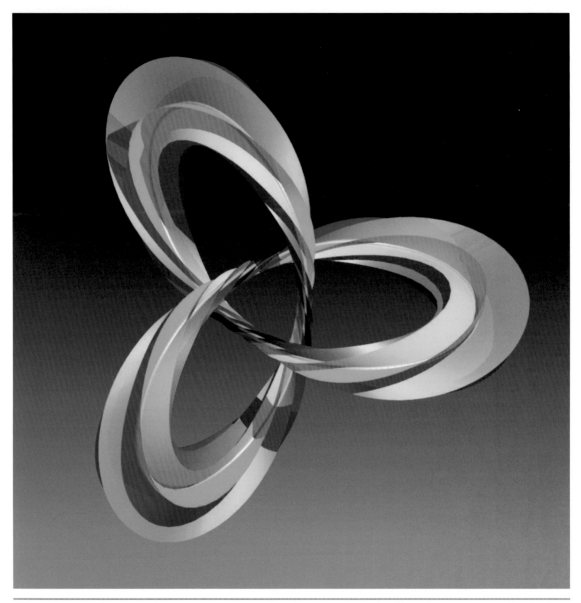

Fibered Knot. *If a knot is fibered, then a "fan" of surfaces can be defined, each one anchored to the knot and collectively filling out all of space. All Lorenz knots are fibered. (Figure courtesy of Jos Leys from the online article, "Lorenz and Modular Flows, a Visual Introduction.")*

A New Twist in Knot Theory

WHETHER YOUR TASTE RUNS to spy novels or Shakespearean plays, you have probably run into the motif of the double identity. Two characters who seem quite different, like Dr. Jekyll and Mr. Hyde, will turn out to be one and the same.

This same kind of "plot twist" seems to work pretty well in mathematics, too. In 2006, Étienne Ghys of the École Normale Supérieure de Lyon revealed a spectacular case of double identity in the subject of knot theory. Ghys showed that two different kinds of knots, which arise in completely separate branches of mathematics and seem to have nothing to do with one another, are actually identical. Every *modular knot* (a curve that is important in number theory) is topologically equivalent to a *Lorenz knot* (a curve that arises in dynamical systems), and vice versa.

The discovery brings together two fields of mathematics that have previously had almost nothing in common, and promises to benefit both of them.

The terminology "modular" refers to a classical and ubiquitous structure in mathematics, the *modular group*. This group consists of all 2×2 matrices, $\begin{bmatrix} a & b \\ c & d \end{bmatrix}$, whose entries are all integers and whose determinant ($ad - bc$) equals 1. Thus, for instance, the matrix $\begin{bmatrix} 2 & 3 \\ 5 & 8 \end{bmatrix}$ is an element of the modular group. (See "Error-Term Roulette and the Sato-Tate Conjecture," on page 19 for another mathematical problem where the concept of modularity is central.)

This algebraic definition of the modular group hides to some extent its true significance, which is that it is the symmetry group of 2-dimensional lattices. You can think of a lattice as an infinitely large wire mesh or screen. The basic screen material that you buy at a hardware store has holes, or unit cells, that are squares. (See Figure 1a, next page.) However, you can create lattices with other shapes by stretching or shearing the material uniformly, so that the unit cells are no longer square. They will become parallelograms, whose sides are two vectors pointing in different directions (traditionally denoted ω_1 and ω_2). The points where the wires intersect form a polka-dot pattern that extends out to infinity. The pattern is given by all linear combinations of the two "basis vectors," $a\omega_1 + b\omega_2$, such that both a and b are integers.

Unlike hardware-store customers, mathematicians consider two lattices to be the same if they form the same pattern of intersection points. (The wires, in other words, are irrelevant, except to the extent that they define where the crossing points are.) This will happen whenever a lattice has basis vectors ω_1' and ω_2' that are linear combinations of ω_1 and ω_2 (i.e., $\omega_1' =$

Étienne Ghys. *(Photo courtesy of Étienne Ghys.)*

$a\omega_1 + b\omega_2$ and $\omega_2' = c\omega_1 + d\omega_2$, for some integers a, b, c, and d) and vice versa. These conditions hold precisely when the matrix $\begin{bmatrix} a & b \\ c & d \end{bmatrix}$ is in the modular group. Figures 1b and 1c show two different bases for a hexagonal lattice. The matrix that transforms one basis into the other would be a member of the modular group.

(a)

(b)

(c)

Figures 1a–1c. *(a) A hardware-store lattice and its basis vectors. (b) A triangular lattice and its basis vectors. (c) The same lattice can be generated by two different vectors. (Figures courtesy of Jos Leys from the online article, "Lorenz and Modular Flows, a Visual Introduction.")*

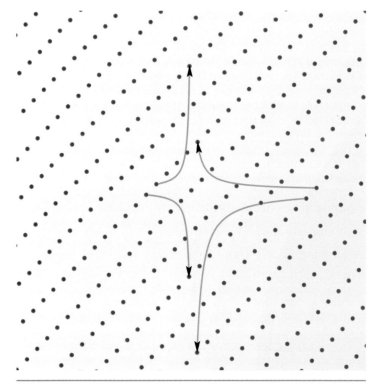

Figure 1d. *The modular flow gradually deforms the shape of a lattice, but brings it back after a finite time to the same lattice with different basis vectors. This figure illustrates the trajectories of four points in the lattice. Every point in the lattice moves simultaneously, and every point arrives at its "destination" in the lattice at the same time. Only the center point does not move at all. (Figure courtesy of Jos Leys from the online article, "Lorenz and Modular Flows, a Visual Introduction.")*

The matrix transformation distorts the underlying geometry of the plane, yet maps the lattice to itself. It accomplishes this transformation in one step. But there is also a way to produce the same effect gradually, by means of a smooth deformation. Imagine drawing a family of hyperbolas, with one hyperbola linking ω_1 to ω_1' and another linking ω_2 to ω_2'. (See Figure 1d.) Remarkably, it is possible to extend this *modular flow* to the entire plane, in such a way that all of the polka dots on the lattice flow along hyperbolas to different polka dots, no polka dots are left out, and the direction of motion of every polka dot at the beginning matches the direction of the new polka dot that comes to replace it at the end of the flow.

There is another way of visualizing the modular flow that emphasizes the special nature of matrices with integer entries. This method involves constructing an abstract "space of all lattices" (which turns out to be three-dimensional, and as described below, can be easily drawn by a computer). The modular flow defines a set of trajectories in this space, in the same way that water flowing in a stream generates a set of streamlines. Most of the streamlines do not form closed loops. However, those that do close up are called *modular knots*, and they correspond explicitly to elements of the modular group. This point of view subtly shifts the emphasis from algebra (the modular group is interesting because its elements have integer entries) to geometry (the modular group is interesting because it produces closed trajectories of the modular flow).

How should we visualize the "space of all lattices"? This turns out to be a crucial question. One traditional approach identifies a lattice with the *ratio* of its two basis vectors, $\tau = \omega_2/\omega_1$. In order for this definition to make sense, the basis vectors have to be considered as complex numbers (i.e., numbers with both a real and imaginary part). The ratio τ will be a complex number $x + iy$ whose imaginary part (y) can be assumed to be positive. Thus τ lies in the upper half of the xy-plane. In fact, it can be pinned down more precisely than that. As explained above, any given lattice can have many different pairs of basis vectors with different ratios τ, but it turns out that there is only one pair of basis vectors whose ratio τ lies in the shaded region of Figure 2 (next page). This "fundamental region" can therefore be thought of as representing the space of all lattices, with each lattice corresponding to one point in the region. For example, the screen you buy in the hardware store, with square holes, corresponds to the ratio $0 + 1i$ or the point $(0, 1)$ in the fundamental region.

However, there is some ambiguity at the edge of the fundamental region. The points τ on the left-hand boundary correspond to the same lattices as the points τ' on the right-hand side. This means that the two sides of the fundamental region should be "glued together" to form a surface that looks like an infinitely long tube with an oddly shaped cap on one end. This two-dimensional surface is called the *modular surface*.

As mentioned above, each lattice is represented by a single point on the modular surface. As the lattice deforms under the modular flow, its corresponding point travels along a circular arc in the upper half-plane. Because of the way the boundaries of the fundamental domain are glued together, each time the curve exits one side of the fundamental domain it re-emerges on the opposite side, and the result is a trajectory with multiple

$$r = e^{2\pi i/3} = -\frac{1}{2} + \frac{\sqrt{3}}{2}i$$

Mapping	Description	Formula	Matrix
X	Rotation about i by π radians	$\dfrac{0z-1}{1z+0}$	$\begin{bmatrix} 0 & -1 \\ 1 & 0 \end{bmatrix}$
Y	Rotation about r by $2\pi/3$ radians	$\dfrac{0z-1}{1z+1}$	$\begin{bmatrix} 0 & -1 \\ 1 & 1 \end{bmatrix}$
Z	Translation by 1 (to the right)	$\dfrac{z+1}{0z+1}$	$\begin{bmatrix} 1 & 1 \\ 0 & 1 \end{bmatrix}$

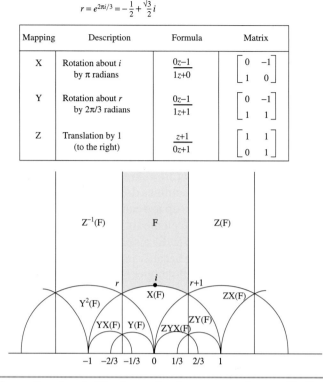

Figure 2. *Any lattice has a basis ω_1, ω_2 whose ratio lies in the fundamental domain F. Any change of basis corresponds to a linear fractional transformation (see column labeled "Formula") or to a matrix (see column labeled "Matrix"). The set of all such transformations is known as the* modular group. *The simple transformations X, Y and Z, listed here, generate the rest of the modular group. Note that the images X(F), Y(F), XY(F), ..., cover a half-plane. The* modular surface *is the quotient space of the half-plane by the modular group. It can be visualized as the region F with its sides glued together. (The bottom is also folded in half and glued together.)*

pieces, somewhat like the path of a billiard ball (see Figure 3). (In fact, the modular flow has sometimes been called *Artin's billiards*, after the German mathematician Emil Artin who studied it in the 1920s.) Most billiard trajectories do not close up, but a few of them do, and these are called *closed geodesics*. They are almost, but not quite, the same as modular knots; the difference is that they lie in a two-dimensional surface, but modular knots are defined in three-dimensional space.

The "missing" dimension arises because there are really four dimensions that describe any lattice: two dimensions of shape, one dimension of orientation, and one of mesh size. The modular surface ignores the last two dimensions. In other words, two lattices correspond to the *same point* in the modular surface if they have the same shape but different orientations. For everyday applications, it makes sense to consider such lattices to be equivalent. If you wanted a screen window made up of diamond shapes instead of square shapes, you wouldn't take your screen back to the store and exchange it; you would simply rotate the lattice 45 degrees.

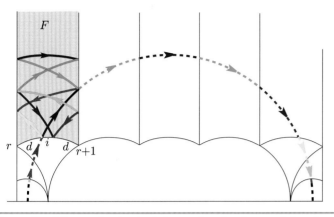

Figure 3. *When a lattice is deformed as by the modular flow, its corresponding basis vectors define a path through the half-plane illustrated in Figure 2. Each segment of this path (indicated here by different colors) may be mapped back to the fundamental region F by a transformation in the modular group.*

However, this common-sense reduction throws away some valuable information about the modular flow. If the orientation is not ignored, the "space of all lattices" becomes three-dimensional.[1] Amazingly, it is *simpler* to visualize this space than it is to visualize the modular surface. Using some elegant formulas from the theory of elliptic curves, Ghys showed that the space of all lattices is topologically equivalent to an ordinary three-dimensional block of wood, with a wormhole bored out of it in the shape of a trefoil knot. (See Figure 4, next page.) Modular knots, therefore, are simply curves in space that avoid passing through the forbidden zone, the trefoil-shaped wormhole in space.

Every matrix in the modular group corresponds to a modular knot, and simpler matrices tend to correspond to simpler knots. For instance, Ghys showed that the matrix $\begin{bmatrix} 1 & 1 \\ 1 & 2 \end{bmatrix}$ corresponds to the green loop shown in Figure 5a, page 9. This loop is unknotted, but it does form a nontrivial link with the forbidden trefoil knot. (That is, it cannot be pulled away from the forbidden zone without passing through it.) The matrix $\begin{bmatrix} 2 & 3 \\ 5 & 8 \end{bmatrix}$ corresponds to a knot that winds around the forbidden zone several times, and is actually a trefoil knot itself. The matrix $\begin{bmatrix} 3997200840707 & 2471345320891 \\ 9088586606886 & 5619191248961 \end{bmatrix}$ corresponds to the white knot in Figure 5b, page 9, which is a bit of a mess. Thus two natural questions arise: What kinds of knots can arise as modular knots? And how many times do they wind around

---

[1]This space should properly be called the space of all *unimodular* lattices because the area of the unit cells is still ignored (or, equivalently, assumed to be equal to 1).

the forbidden zone? Remarkably, Ghys answered both of these questions. But the answer requires a detour into a completely different area of mathematics.

Figure 4. *Ghys realized that the conventional approach to defining the modular surface omits information about the orientation of a lattice. Therefore he defined a modular space, which is topologically equivalent to the exterior of a trefoil knot, as shown here. Closed geodesics, like the one in Figure 3, lift to modular knots, which wind around the trefoil but never intersect it. (Figure courtesy of Jos Leys from the online article, "Lorenz and Modular Flows, a Visual Introduction.")*

In 1963, a mathematician and meteorologist named Edward Lorenz was looking for a simple model of convection in the atmosphere, and came up with a set of differential equations that have become iconic in the field of dynamical systems. The equations are these: $\frac{dx}{dt} = 10(y - x)$, $\frac{dy}{dt} = 28x - y - xz$, $\frac{dz}{dt} = xy - \frac{8}{3}z$.

The specific meaning of the variables x, y, z is not too important. They are three linked variables, each a function of time (t), which in Lorenz's model represented the temperature and amount of convection at time t in a fictitious atmosphere. The equations describe how the atmosphere evolves over time. Because there are only three variables, unlike the millions of variables necessary to describe the real atmosphere, the solutions can be plotted easily as trajectories in three-dimensional space.

Lorenz noticed a phenomenon that is now known as deterministic chaos or the "butterfly effect." Even though the equations are completely deterministic—there is no randomness in this fictitious atmosphere—nevertheless it is impossible to forecast the weather forever. No matter what

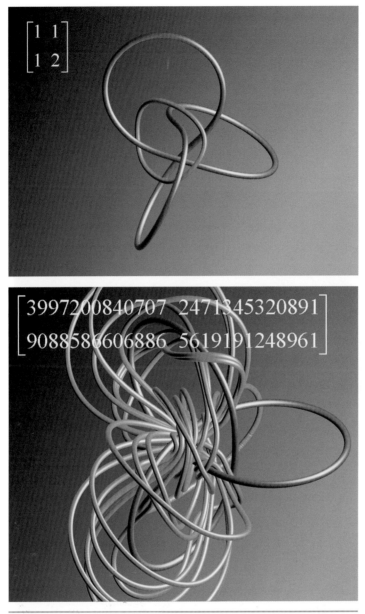

Figure 5. *Two different modular knots and the corresponding elements of the modular group. (Figures courtesy of Jos Leys from the online article, "Lorenz and Modular Flows, a Visual Introduction.")*

starting point (x, y, z) you choose, even the slightest deviation from this initial condition corresponding to a slight experimental error in measuring the temperature or convection) will eventually lead to completely different weather conditions. The name "butterfly effect" refers to an often-cited analogy: the flapping of a butterfly's wings today in Borneo could lead to a typhoon next month in Japan.

A look at the trajectories of the Lorenz equation (Figure 6) explains why this is so. The trajectories concentrate around two broad, roughly circular tracks that, ironically, bear some resemblance to a pair of butterfly wings. You can think of one loop as predicting dry, cold weather and the other as predicting rainy, warm weather. Each time the trajectory circles one ring of the track (one "day"), it returns to the intricately interwoven region in the center, where it "decides" which way to go the next day. After a few dozen circuits, all information about the starting position is effectively lost, and the trajectory might as well be picking its direction at random.[2] Thus the full trajectory is, for all practical purposes, unknowable.

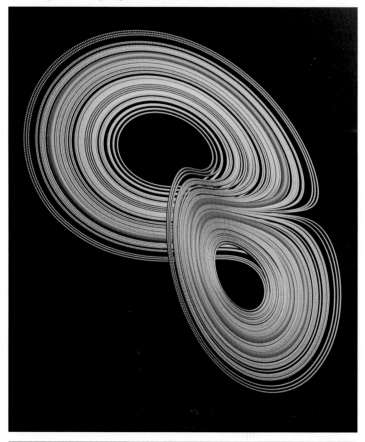

Figure 6. *The Lorenz attractor (yellow). One particular trajectory is shown in blue. It is a closed orbit of the Lorenz differential equations, or a* Lorenz knot. *(Figure courtesy of Jos Leys from the online article, "Lorenz and Modular Flows, a Visual Introduction.")*

However, suppose we aren't being practical. Suppose we can prescribe the initial position with unrealistic, infinite precision.

---

[2]Incidentally, Lorenz chose the coefficient 28, in the equation for dy/dt, to postpone the onset of chaos as long as possible. A trajectory starting near the origin $(0, 0, 0)$ will loop around one track of the Lorenz attractor 24 times before it finally switches to the other side. If the coefficient 28 was made either larger or smaller, the number of consecutive "sunny days" at the beginning of the trajectory would decrease.

Can we find a trajectory of the Lorenz equations that is the epitome of predictability—a closed loop? If so, the weather on day 1 would be repeated exactly on day 7, on day 13, and so on. Order, not chaos, would reign in the world.

This was the question that Bob Williams (now retired from the University of Texas) began asking in the late 1970s, together with Joan Birman of Columbia University. Over a thirty-year period, it has gradually become clear that closed trajectories do exist and that they form a variety of nontrivial knots. It is natural to call them *Lorenz knots.* For example, the trajectory shown in Figure 7a is topologically equivalent to a trefoil knot.

Even though perfectly periodic weather conditions can never be realized in practice, nevertheless they "give you a feeling of how tangled up this flow is," Birman says. It is amazing to discover that the simple pair of butterfly wings seen in Figure 6 actually contains infinitely many different Lorenz knots, all seamlessly interwoven without ever intersecting one another. (See Figure 7b.) However, there are also many knots that do *not* show up as trajectories of the Lorenz equations. For example, the second-simplest nontrivial knot, the figure-eight knot (Figure 7c) is *not* a Lorenz knot. (The simplest nontrivial knot, the trefoil knot, is a Lorenz knot because we have seen it in Figure 7a.)

In a paper written in 1982, Birman and Williams derived a host of criteria that a knot must satisfy in order to be a Lorenz knot. For example, they are *fibered* knots—an extremely unusual property that means that it is possible to fill out the rest of space with surfaces whose boundaries all lie on the knot. The figure on page 3, "The Fibered Knot," illustrates this difficult-to-visualize property. Using Birman and Williams' criteria, Ghys has showed that only eight of the 250 knots with ten or fewer "overpasses" or "underpasses" are Lorenz knots. In other words, even though infinitely many different Lorenz knots exist, they are rather uncommon in the universe of all knots.

Lorenz knots are extremely difficult to draw because the very nature of chaos conspires against any computer rendering program. Even the slightest roundoff error makes the knot fail to close up, and eventually it turns into a chaotic tangle. Thus, without theoretical results to back them up, we would not know that computer-generated pictures like Figure 7a represent actual closed trajectories.

Birman and Williams originally proved their theorems about Lorenz knots under one assumption. Earlier, Williams had constructed a figure-eight-shaped surface, a *geometric template,* which (he believed) encoded all the dynamics of a Lorenz knot. In essence, he argued that the butterfly wings of Figure 6 are real (see also Figure 8, p. 12), and not just a trick of the eye. Although he and Clark Robinson of Northwestern University produced strong numerical evidence to support this belief, a proof remained elusive. In fact, this problem appeared on a list of leading "problems for the 21st century" compiled by Stephen Smale in 1998.

As it turned out, Smale (and Birman and Williams) did not have long to wait. In 2002, Warwick Tucker of Uppsala University proved the conjecture by using interval arithmetic, a hybrid

(a)

(b)

(c)

Figure 7. *(a) This Lorenz knot is topologically equivalent to a trefoil knot. (b) Different Lorenz knots interlace with each other in a phenomenally complex way, never intersecting one another. (c) Not all topological knots are Lorenz knots. For example, no orbit of the Lorenz equations is topologically equivalent to a figure-eight knot, shown here. (Figures courtesy of Jos Leys from the online article, "Lorenz and Modular Flows, a Visual Introduction.")*

technique that combines computer calculations with rigorous proofs that the results are robust under roundoff error.

Even though the actual trajectories of the Lorenz flow do not lie on the template, Tucker's work guarantees that they can be mashed down onto the template without altering the topological type of the knot. In other words, none of the strands of the knot pass through each other or land on top of one another as a result of the mashing process. Thus, the topology of Lorenz knots can be studied simply by drawing curves on the template, which is a much easier job than solving Lorenz's equations.

Once the knots have been pressed onto the template, the shape of the knot is determined by the series of choices the trajectory makes as it passes through the central region. Each time it chooses to veer either left or right. The trefoil knot in Figure 7a would correspond to the string of decisions "right, left, right, left, right," or simply the string of letters RLRLR. (Different ways of reducing the three-dimensional dynamics to a one-dimensional "return map" had been noted by other mathematicians too, including John Guckenheimer of Cornell University.)

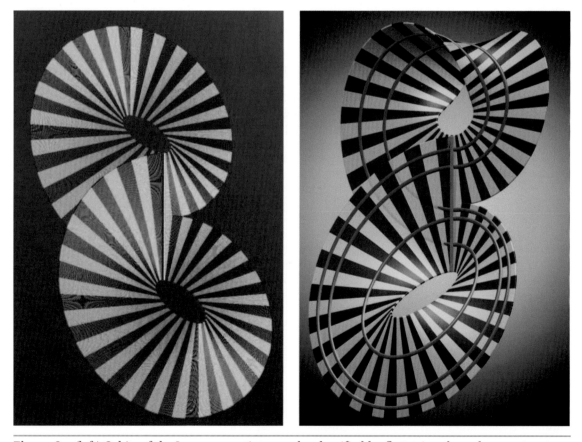

Figure 8. *(left) Orbits of the Lorenz equations can be classified by flattening them down onto a* template, *a sort of paper-and-scissors model of the Lorenz attractor. (right) A related set of equations, Ghrist's dynamical system, has a template that looks like the Lorenz template with an extra half-twist to one of the lobes. Amazingly, this slight modification is enough to guarantee that* every *knot appears as a closed orbit. (Figures courtesy of Jos Leys from the online article, "Lorenz and Modular Flows, a Visual Introduction.")*

The idea of templates has led to other surprising discoveries in dynamical systems. In a 1995 doctoral dissertation, Robert Ghrist, who is now at the University of Illinois, showed that some dynamical systems allow a much richer set of closed trajectories—in fact, they contain *every* closed knot. Not just every known knot, but every knot that will ever be discovered! Ghrist points out that his example can be made quite concrete. "Suppose I take a loop of wire," he says, "and bend it in the shape of a figure-eight knot [Figure 7c]. I run an electric current through it, and look at the induced magnetic field. Assuming it doesn't have any singularities, it will have closed field lines of all knot types." Once again, the proof involved the construction of a geometric template, which turned out to be very similar to the butterfly-shaped Lorenz template, except that one of the lobes is given an extra twist. (See Figure 8.)

Bob Williams. *(Photo courtesy of R. F. Williams.)*

Although Ghrist's result surprised Birman and Williams, who had conjectured that a "universal template" containing all knots was impossible, it confirmed the idea that some flows are more chaotic than others, and that studying closed trajectories is a good way of telling them apart. "With the wisdom of hindsight, the existence of a small number of knots in a flow is like the onset of chaos," Birman says. Lorenz's equations seem to define a relatively mild form of chaos, while Ghrist's equations seem to represent a full-blown case.

When Ghys started thinking about modular knots, he ironically made the opposite guess to Birman and Williams. He thought that the class of modular knots was probably universal—in other words, that every knot can be found somewhere in the modular flow. "The first time I thought of these questions, I wanted to understand modular knots, and I had no idea they were connected with the Lorenz equations," Ghys says. But then he made a remarkable discovery. There is a copy of the Lorenz template hidden within the modular flow! He originally made a schematic drawing that shows the template looking very much like a pair of spectacles straddling the trefoil-shaped "forbidden zone." Later, with the help of graphic artist Jos Leys of Belgium, he produced beautiful animations that show how any modular knot can be deformed onto the template. (See Figure 9, next page.) "A proof for me is not always fully formalized," he says. "I had it all clear in my head. I was sure it was true, I knew why it was true, and I was beginning to write it down. But for me, putting it in a picture was a confirmation that it was more true than I thought."

Whether by formal argument or a "proof by picture," the conclusion immediately follows that all modular knots are Lorenz knots. "A lot of people look at Étienne's result and see it as a result about how complicated the modular flow is, or the number theory," says Ghrist. "I take a contrarian view. Étienne's work is showing that there is a certain parsimony in modular flow. Working under the constraints of complicated dynamics, it's got the simplest representation possible. Likewise, with the Lorenz equations being one of the first dynamical systems investigated, it's not surprising that they also have the simplest kind of chaotic dynamics."

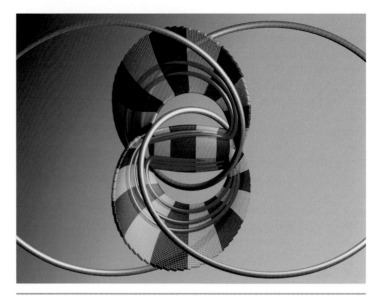

Figure 9. *The Lorenz template can be deformed in such a way that it straddles the "forbidden" trefoil from Figure 4. This insight is a key part of the proof that any Lorenz knot is a modular knot, and vice versa. (Figure courtesy of Jos Leys from the online article, "Lorenz and Modular Flows, a Visual Introduction.")*

The converse—that all Lorenz knots are modular—also holds true, once it is shown that the modular group allows all possible sequences of "left turns" and "right turns" within the template. In fact, Ghys found a direct connection between the sequence of turns and a previously known function in number theory. Ghys calls it the "Rademacher function," although he comments that so many mathematicians have discovered and rediscovered it that he is not sure whether to name it after "Arnold, Atiyah, Brooks, Dedekind, Dupont, Euler, Guichardet, Hirzebruch, Kashiwara, Leray, Lion, Maslov, Meyer, Rademacher, Souriau, Vergne, [or] Wigner"! The Rademacher function assigns to each matrix $\begin{bmatrix} a & b \\ c & d \end{bmatrix}$ in the modular group an integer. The classical, and not very intuitive way, of defining this integer goes as follows: First you sum two nested infinite series of complex numbers that are not integers. After computing the sums (called g_2 and g_3), next you compute $(g_2)^3 - 27(g_3)^2$ (the "Weierstrass discriminant"), and take its 24th root (the "Dedekind eta function"). Finally, you take the complex logarithm of this function. It is well known that complex logarithms have an ambiguity that is an integer multiple of $2\pi i$. When you traverse the closed geodesic defined by $\begin{bmatrix} a & b \\ c & d \end{bmatrix}$ and come back to the starting point, the logarithm of the Dedekind eta function will not necessarily come back to its original value. It will change by $2\pi i$ times an integer—and that integer is the Rademacher function of $\begin{bmatrix} a & b \\ c & d \end{bmatrix}$.

A variety of other ways to compute this function were known, but none of them can be said to be really easy. Ghys' work gives it a new topological interpretation that does not require such an elaborate definition and that makes its meaning completely clear. As explained above, the idea is to use modular knots in three-space instead of closed geodesics in the modular surface. Every matrix in the modular group defines a modular knot. From Ghys's work, it follows that you can press this knot down onto the Lorenz template. Then the Rademacher function is simply the number of left turns minus the number of right turns! It's hard to imagine a more elegant or a more concrete description.

Ghys's result opens up new possibilities both for number theory and for dynamical systems. One reason mathematicians care so much about two-dimensional lattices is that they are the next step up from one-dimensional lattices. In one dimension, up to scaling, there is only one lattice: the set of integers, $\{\ldots, -2, -1, 0, 1, 2, \ldots\}$. The discipline of number theory (properties of the integers) is an exploration of its properties.

In fact, there is a precise analogy between number theory and the modular surface, which has yet to be fully understood. One of the most important functions in number theory is the Riemann zeta function $\zeta(s)$, which relates the distribution of prime numbers to the distribution of a mysterious set of non-integers, the points s_n where $\zeta(s_n) = 0$. These points s_n are known as zeroes of the zeta function.

The most famous open problem in number theory, the Riemann Hypothesis (see *What's Happening in the Mathematical Sciences*, Volumes 4 and 5), asks for a proof that the numbers s_n all lie on a single line in the complex plane. This kind of tight control over their distribution would imply a host of "best possible" results about the distribution of prime numbers.

One of the many pieces of evidence in favor of the Riemann Hypothesis is a very similar theorem for two-dimensional lattices, called the *Selberg trace formula*, that was proved in 1956 by Atle Selberg (a Norwegian mathematician who died recently, in 2007). It involves a *Selberg zeta function*, whose zeroes can be described as the energy levels of waves on the modular surface, and whose formula looks eerily similar to the formula for the Riemann zeta function. And what plays the role of prime numbers in that formula? The answer is: the lengths of closed geodesics in the modular surface. To make a long story short, the Selberg trace formula says that these lengths are dual to energy levels of waves on the modular surface, in exactly the same way that prime numbers are thought to be dual to zeroes of the Riemann zeta function. In fact, this analogy has led some mathematicians and physicists to suggest that the Riemann zeroes may also turn out to be energy levels of some yet undiscovered quantum-mechanical oscillator.

In any event, closed geodesics on the modular surface are clearly very relevant to number theory. And Ghys' result suggests that there is much more information to be obtained by going up a dimension and looking at modular knots. The Rademacher function is only the tip of the iceberg. It represents the simplest topological invariant of a modular knot, namely the "linking number," which describes how many times it wraps around the forbidden trefoil. Knot theory offers many

The Rademacher function is only the tip of the iceberg. It represents the simplest topological invariant of a modular knot, namely the "linking number," which describes how many times it wraps around the forbidden trefoil.

more possible invariants for a modular knot. Could some of these also have analogues in number theory?

"I must say I have thought about many aspects of these closed geodesics, but it had never crossed my mind to ask what knots are produced," says Peter Sarnak, a number theorist at Princeton University. "By asking the question and by giving such nice answers, Ghys has opened a new direction of investigation which will be explored much further with good effect."

Ghys' theorem also implies that modular knots, because they are Lorenz knots, have the same properties that Lorenz knots do. For instance, modular knots are fibered—a fact that was not previously known. Ghys is currently looking for a more direct proof of this fact.

The double identity of modular and Lorenz knots also raises new questions for dynamical systems. For starters, modular knots are vastly easier to generate than Lorenz knots because the trajectories are parametrized by explicit functions. These trajectories are not literally solutions of the Lorenz flow, and yet somehow they capture an important part, perhaps all, of its dynamical properties. How faithfully does the modular flow really reflect the Lorenz flow? Tali Pinsky of Technion in Israel recently showed that there is a trefoil knot in space that forms a "forbidden zone" for the Lorenz flow, analogous to the forbidden zone for the modular flow. More generally, what kinds of dynamical systems have templates, and when is a template for a solution just as good as the solution itself? How can you tell whether a template is relatively restrictive, like Lorenz's, or allows for lots of different behaviors, like Ghrist's?

Ghys' theorem has already inspired Birman to take a fresh look at Lorenz knots. With Ilya Kofman of the College of Staten Island, she has recently come up with a complete topological description of them. It was already known that all torus knots—in other words, curves that can be drawn on the surface of a torus—are Lorenz knots. (For example, a trefoil knot can be drawn as a curve that goes around the torus three times in the "short" direction while going twice around in the "long" direction.) However, the converse is not true—many Lorenz knots are not torus knots.

Birman and Kofman have shown that Lorenz knots are nevertheless related to torus knots by a simple twisting procedure. The idea of twisting is to take several consecutive strands of the knot—any number you want—and pull them tight so that they lie parallel to each other. Cut all of the parallel pieces at the top and bottom, to get a skein. Now give the whole skein a twist by a rational multiple of $360°$, so that the bottom ends once again lie directly below the top ends. Then sew them back up to the main knot exactly where they were cut off in the first step.

Any Lorenz knot, Birman and Kofman showed, can be obtained from a torus knot by repeated application of this twisting procedure.[3]

In fact, topologists were already aware that twisted torus knots have some special properties. In 1999, topologists Patrick Callahan, John Dean, and Jeff Weeks proved that their Jones polynomials (a knot invariant discovered in the 1980s by Vaughan Jones) are unusually simple. Another important topological invariant of knots, discovered by William Thurston in the 1970s, is the hyperbolic volume of their complement. Callahan, Dean and Weeks showed that twisted torus knots tend to have unusually small volumes. Their results, together with the work of Ghys and Birman, suggest that twisted torus knots arise naturally in problems outside topology because they are the simplest, most fundamental non-torus knots.

"What I'm most pleased about is that Ghys' work is reminding a new generation of mathematicians of what Joan Birman and Bob Williams did back in the 1980s," Ghrist says. "That was incredibly beautiful and visionary work that they did. I'm delighted to see someone of Ghys' stature and talent coming in, revisiting those ideas and finding new things."

Joan Birman. *(Photo courtesy of Joan S. Birman.)*

---

[3]There is one mild technical condition, which is that the twists all have to be in the same direction.

Schwarz Reflection. *The Schwarz reflection principle is a simple way of extending an analytic function (here graphed in blue) to a region where it was not originally defined (orange). L-functions of elliptic curves can likewise be extended by using their symmetry properties. (Figure courtesy of Andrew D. Hwang.)*

Error-term Roulette and the Sato-Tate Conjecture

A T FIRST BLUSH, NUMBER THEORY—the study of the properties of whole numbers—seems as if it should be the most exact branch of mathematics. After all, when you factor an integer (say, $6 = 2 \times 3$), this is a precise statement, with no errors or approximations involved.

Yet on a deeper analysis, integers often act as if they obeyed probabilistic laws. The simplest example of this phenomenon is the Prime Number Theorem, proved in 1898, which (roughly speaking) says that the "probability" of a number being prime is inversely proportional to the number of digits in the number. More precisely, the Prime Number Theorem computes the average number of primes you would expect to find in a randomly chosen interval. (See Table 1.)

However, you cannot understand a random process with averages alone. The very essence of randomness is that things do not always behave in the average way. Sometimes there will be more primes than expected in a given interval, sometimes fewer (as shown in Table 1). In any science, measurements that are subject to random error are normally presented with a confidence interval—in other words, an estimate of the *distribution* of these random deviations from the mean. So it would be only natural for number theorists to aspire to the same thing.

n	Interval	Number of primes in intervals	p(n)	1/ln(n)	ratio of exact to approximate probability
100	[100, 110)	4	.4	.217417	1.8421
10,000	[10,000, 10,100)	11	.11	.108574	1.01314
100,000	[100,000, 101,000)	75	.075	.0723824	1.03616
1,000,000	[1,000,000, 1,010,000)	551	.0551	.0542868	1.01498
10,000,000	[10,000,000, 10,100,000)	4306	.04306	.0434294	.991493
100,000,000	[100,000,000, 101,000,000)	36249	.036249	.0361912	1.0016

Table 1. *The prime number theorem (1898) was the first indication that prime numbers behave as if they were generated by a random process. If a number is chosen randomly from an interval that begins at n, the probability P(n) that it is prime [column 4] is roughly equal to 1/ln(n) [column 5].*

In the example just cited—the problem of counting prime numbers in an interval—number theorists do not yet know the correct distribution of deviations from the mean. They only have a conjecture, known as the Riemann Hypothesis. (See "Think and Grow Rich," *What's Happening in the Mathematical*

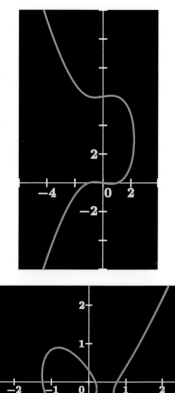

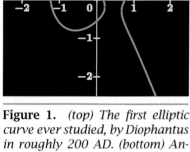

Figure 1. *(top) The first elliptic curve ever studied, by Diophantus in roughly 200 AD. (bottom) Another typical elliptic curve. Note that elliptic curves over the real numbers may have either one or two components. (Figures courtesy of Andrew D. Hwang.)*

Sciences, Volume 5.) However, for some other number-theoretic questions, which are not too distantly related to the Riemann Hypothesis, the distribution of deviations from the mean are now known. In the game of "error-term roulette," as Harvard mathematician Barry Mazur calls it, mathematicians can now come to the gaming table armed with some knowledge of how the bouncing ball behaves.

In 2006, a group of four mathematicians solved a 40-year-old problem called the Sato-Tate Conjecture. Like the Riemann Hypothesis, this conjecture deals with a counting problem. In this case, the items to be counted are points on an elliptic curve. As in the Riemann Hypothesis, the "average" behavior of these counts has been well-known for decades. The Sato-Tate Conjecture, like the Riemann Hypothesis, posits a specific distribution for the deviations from the average. But unlike the Riemann Hypothesis, the Sato-Tate Conjecture is now a proven fact.

To understand what the Sato-Tate Conjecture means, one first needs to understand elliptic curves. As generations of students have learned to their dismay and confusion, elliptic curves are *not* ellipses, and in the context of greatest interest to number theorists, they are not even curves. They are, to be precise, the solution sets to a nonsingular cubic equation in two variables. Such equations arise, for example, in finding the arc-length parametrization of an ellipse. (This accounts for the very tenuous connection between elliptic curves and ellipses). Elliptic curves also arise in a variety of classic number theory problems.

The oldest known problem that leads to an elliptic curve was posed by Diophantus in his book *Arithmetica*, around the year 250:

"To divide a given number into two numbers such that their product is a cube minus its side."

What does this mean? If the "given number" is denoted a, then Diophantus is looking for two numbers whose sum is a. They can be denoted y and $a - y$. Then their product is $y(a - y)$; and Diophantus wants this number to be expressible as the volume of a cube (x^3) minus its side length (x). Thus, his problem would nowadays be written $y(a - y) = x^3 - x$. Diophantus did not provide a general procedure for solving this problem, but in the case $a = 6$ he derived the not-so-obvious solution, $y = 26/27$ and $x = 17/9$.

In all of his problems Diophantus assumed the solutions to be rational numbers. This was a very important restriction because the nature of an elliptic curve depends a great deal on the universe of potential solutions you allow for x and y. When x and y are both assumed to be real numbers, there are infinitely many solutions. The graph of an elliptic curve, in this universe, does in fact look like a smooth curve, which can have either one or two pieces. (See Figure 1.) In some cases, the "neck" between the two pieces may collapse down to a single point, creating a curve with a singularity (Figure 2), and this technically disqualifies the curve from being elliptic.

Other solution sets (or "fields") besides the real numbers are also possible, and in those fields the elliptic curve bears no resemblance to a curve. For instance, when x and y are allowed to be complex numbers, the elliptic curve becomes a surface. Surprisingly, no matter what cubic equation you look at, the surface always has basically the same shape—a torus. The real-number solutions form a curve embedded in the torus.

Ever since Diophantus, number theorists have been most interested in rational solutions to elliptic curves. Thus they assume that x and y lie in the field of rational numbers. Over the rational numbers, different cubic equations can produce elliptic curves with very different "shapes." Sometimes the curve may turn into just a finite set of points. For example, the equation $y^2 = x^3 - x$ (only a slight modification of Diophantus' equation) has only the three "trivial" solutions $(x, y) = (0, 0), (1, 0), (-1, 0)$. In other cases, such as the one considered by Diophantus, there are infinitely many rational solutions.

Modern number theorists have at their disposal many tools for studying elliptic curves that Diophantus did not know about. One of their tricks is to study the equation *modulo a prime number p*. This means that x and y are assumed to lie in the finite field, $F_p = \{0, 1, 2, \ldots p - 1\}$. (Addition and multiplication in this field are defined by taking remainders on division by p. Thus, for example, in the finite field F_3, we have $2 + 2 = 1$, because the remainder of $2 + 2$ after division by 3 is 1.) The advantage of finite fields is that they break up the "global" problem of solving the cubic equation into many smaller, "local" pieces. Because there are only finitely many possibilities for x and y, all of the solutions can be found simply by trial and error. And if the number of solutions to an equation modulo p should happen to be zero, it usually follows that the number of rational-number solutions is also zero.

The Sato-Tate Conjecture, then, pertains to the number of solutions an elliptic curve has modulo any prime number p. The average behavior has been understood since the 1930s, when Helmut Hasse proved that the number of solutions is approximately $p + 1$. Hasse's theorem means that the number of solutions is roughly (asymptotically) the same as the number of solutions to a linear equation.

Of course, the devil is in the details: What does "approximately" mean? Hasse gave a very good answer. He showed that for all except a finite number of primes p, the deviations from the mean are less than $2\sqrt{p}$.

In the early 1960s, Michio Sato, a mathematician at the Institute for Advanced Study in Princeton, began tabulating the deviations from the mean using a computer. Exactly what Sato did is "shrouded in mystery" because it was never published, according to Nicholas Katz of Princeton University. Presumably, Sato saw that the deviations follow a specific distribution. Surprisingly, it is quite different from the familiar bell-shaped curve of probability theory; in fact, it is an ellipse. The frequency distribution of the errors seems to cluster around the curve $y = \frac{2}{\pi}\sqrt{1 - x^2}$. In 1963 John Tate of Harvard University, having heard of Sato's numerical experiments, wrote this down as a formal conjecture.

Although Sato's data were unpublished, Figure 3 (see next page) shows what he might have seen. The table lists data for five different elliptic curves, which are labeled 11a, 36a, 96b, 198d, and 8732a according to a standard identification system used in the field. (In fact, 8732a is the same curve that Diophantus examined.) The errors have been normalized to lie between

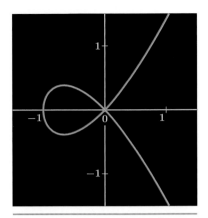

Figure 2. *Cubic curves with a singularity (such as the crossing point in this picture) are usually not considered to be elliptic. (Figure courtesy of Andrew D. Hwang.)*

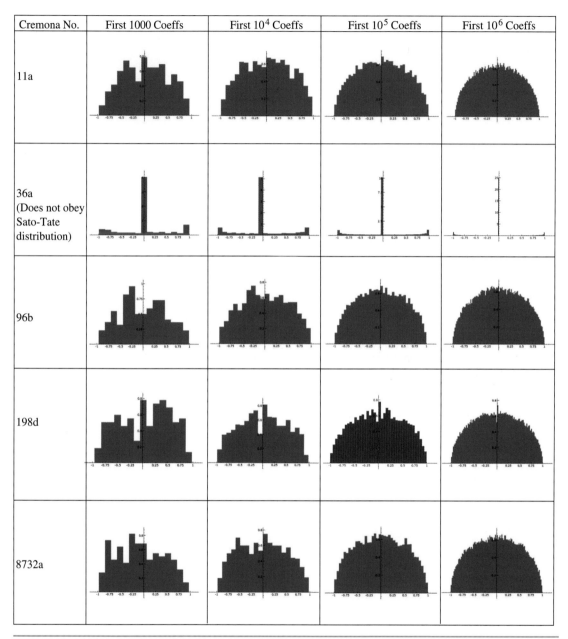

Figure 3. *Five elliptic curves and the histograms of the normalized errors, $a_p = (p + 1 - N_p)/2\sqrt{p}$. (Figure courtesy of William Stein. The plots were done using the free open source program Sage (http://www.sagemath.org).)*

−1 and 1 (rather than between $-2\sqrt{p}$ and $2\sqrt{p}$), and then tabulated in a histogram. In the second column the prime numbers p range from 1 to 1000; in the third column they range from 1 to 10,000; and so on. In four of the five cases, the histograms seem to be approaching the same ellipse, the Sato-Tate distribution. Elliptic curve 36a belongs to an unusual class of curves with an extra kind of symmetry known as "complex multiplication." The errors for these curves do not follow the Sato-Tate distribution; in fact, a large proportion of the errors are 0, so the limiting distribution in this case has an extremely sharp central peak, known as a delta function. Thus the Sato-Tate conjecture applies only to elliptic curves without complex multiplication.

"When [Tate] made the Sato-Tate Conjecture, he had good theoretical reasons for believing it," says Richard Taylor of Harvard. "Atypically for Tate, he didn't write these reasons very clearly. Usually he wrote with beautiful clarity, but when I read this, it's as if he hadn't quite thought it through. Then Serre wrote it very clearly in his book, *Abelian l-Adic Representations and Elliptic Curves* [published in 1968]. At that point it became clear that what you had to do was take L-functions of symmetric powers and prove analytic properties about them."

Jean-Pierre Serre, who had become in 1954 the youngest winner ever of the Fields Medal, was based at the Collège de France in Paris, but paid many visits to Harvard and Princeton and thus was well acquainted with the work of Sato and Tate. His book laid out a clear plausibility argument for the Sato-Tate Conjecture, which was not a proof but a road map for how a proof should go.

Taylor, along with his collaborators Laurent Clozel, Nicholas Shepherd-Barron, and Michael Harris, have now completed the journey outlined by Serre's road map, but it has certainly taken them in some directions that Serre never anticipated. As Mazur describes it, "The Sato-Tate Conjecture comes from a merging of three theories that come together just right—better than you would expect. It's like continental drift in reverse." The three theories are L-functions, automorphic forms, and Galois representations.

While Mazur's continental-drift analogy is appealing, it might be best to think of the theories as machines with three different purposes. L-functions *encode* number-theoretic information, such as the distribution of prime numbers or the number of solutions to elliptic curves. The theory of automorphic forms *extends* the L-function of the elliptic curve past the region where it is originally defined, and thus makes it possible to decode information from the L-function that was previously hidden. Finally, Galois representations *link* the two other machines together. Specifically, they link the local symmetries of an elliptic curve with a global symmetry of its L-function called *modularity*, which turns it into an automorphic form, and thus allows the second machine to work on it. This step can be described in a simple phrase, coined by Kenneth Ribet of the University of California at Berkeley: "Modularity is contagious." Of these three machines, Galois representation theory is the one that has seen the most rapid improvement in recent years. Those improvements—due in large measure to Taylor and even more to his mentor, Andrew Wiles—made the proof of the Sato-Tate Conjecture feasible.

Richard Taylor. *(Photo courtesy of Richard Taylor.)*

Michael Harris. *(Photo courtesy of Michael Harris.)*

Laurent Clozel.

The first of these three giant machines is the most concrete, and yet perhaps the most mysterious for people who have not studied number theory deeply.

Remember that the Sato-Tate Conjecture deals with the difference between the predicted number of solutions to an elliptic curve modulo p and the actual number, N_p. The difference is simply $(p + 1 - N_p)$, and for convenience it is usually denoted by a_p. By Hasse's theorem, a_p always lies between $-2\sqrt{p}$ and $2\sqrt{p}$, and hence $a_p = 2\cos\theta_p\sqrt{p}$, for some angle θ_p. Though it is simple trigonometry, this observation made by Hasse has amazingly deep consequences.

The goal of the Sato-Tate Conjecture is to prove that the numbers a_p, and therefore the angles θ_p, follow a specific probability distribution. This distribution can be rephrased (roughly) as follows: The probability that θ_p equals a given angle θ is proportional to $\sin^2\theta$. Thus the "most likely" values of θ_p are near $\pm\pi/2$, where $\sin^2\theta_p$ is near 1, $\cos\theta_p$ is near 0, and a_p is near 0. Note that this is completely consistent with Figure 3. Similarly, the "least likely" values for a_p are near $\pm 2\sqrt{p}$, where $\cos\theta_p$ is near ± 1 and $\sin^2\theta_p$ is near 0.

Instead of saying that the probability distribution is proportional to $\sin^2\theta$, it is convenient to say instead that the values of θ_p are "equidistributed" with respect to the measure $\sin^2\theta d\theta$. According to Nicholas Katz of Princeton University, you can think of L-functions as a machine for proving equidistribution theorems. That is, in fact, how they made their first appearance in mathematics, when Gustav Lejeune Dirichlet used them to prove that prime numbers are "equidistributed" among all arithmetic progressions. (See Box, "Dirichlet and the Distribution of Prime Numbers".)

Dirichlet's proof had two key ingredients—the "analytic properties" about the L-function that Taylor referred to in his quote above. First, Dirichlet had to show that the L-function could be defined on the line $x = 1$ (in fact, it can be "analytically continued" past that line, and defined for any complex number s); second, he had to show that the value of the L-function at the point $(1, 0)$ [or, in terms of complex numbers, at the point $s = 1 + 0i$] is nonzero. In fact, his original result was not quite strong enough to imply equidistribution in the currently accepted sense; for this one needs the stronger result that the L-function is nonzero along the entire line $x = 1$.

All of these behaviors of L-functions are by now very familiar to number theorists. The original definition in terms of a series or product rarely converges at the most "interesting" points; the boundary between convergence and non-convergence is always a vertical line; the L-function must always be extended in some way past this line; and the farther to the left its zeroes (that is, the points where $L(s) = 0$) can be pushed, the more precise information it provides about the statistical distribution of number-theoretic quantities.

Now, after this extended digression into nineteenth-century math, we return to the twentieth century. To prove any kind of equidistribution result about the numbers a_p—or, equivalently, about the angles θ_p—one needs two ingredients, a character function (see Box, "Dirichlet and the Distribution of Prime Numbers) and an L-function to package it in. Hasse and the French mathematician André Weil realized that there

Dirichlet and the Distribution of Prime Numbers

In the 1830s, Gustav Lejeune Dirichlet introduced what are now called the Dirichlet L-functions,

$$L(\chi, s) = \sum_{n=1}^{\infty} \frac{\chi(n)}{n^s}.$$

Here s, the variable, represents an arbitrary complex number, and $\chi(n)$ represents a *Dirichlet character*. In other words, χ is a function from the integers to the complex numbers that obeys a nice multiplicativity property ($\chi(mn) = \chi(m)\chi(n)$) and is constant on certain arithmetic progressions (that is, there is some N such that for any m, $\chi(m) = \chi(m + N) = \chi(m + 2N)$ and so on). Like all other L-functions, Dirichlet L-functions can also be represented as an infinite *product*:

$$L(\chi, s) = \prod_p \frac{1}{1 - \chi(p)p^{-s}}.$$

The product ranges over all primes p—not over all integers. The identity can be proved by a beautiful trick that was essentially discovered by Leonhard Euler in the 1700s: Write each fraction as the sum of a geometric series, and then expand the product of all these infinitely many geometric series term-by-term. However, to make this trick pass modern mathematical standards of rigor, one must require the series and the product to be absolutely convergent, which is only true if s lies to the right of the line $x = 1$ in the complex plane.

Notice that in the product form of the Dirichlet L-function, one only needs to know the value of the character χ at prime numbers, and this value is constant on arithmetic progressions. Using these facts, Dirichlet was able to show that prime numbers are equidistributed among the different arithmetic progressions modulo N. For instance, taking $N = 10$, there are "equally many" primes in the four different arithmetic progressions $\{1, 11, 21, 31, \ldots\}$, $\{3, 13, 23, 33, \ldots\}$, $\{7, 17, 27, 37, \ldots\}$, and $\{9, 19, 29, 39, \ldots\}$. Or to put it another way, a number that ends in 1 is just as likely to be prime as a number that ends in 3, a number that ends in 7, or a number that ends in 9. Dirichlet's theorem may seem "intuitively obvious," but it was the first theorem of its type, a statement about primes that can be interpreted probabilistically. Without L-functions, we would probably still be looking for a proof.

are actually two "correct" character functions. They are the functions $p \to \sqrt{p}\exp(i\theta_p)$ and $p \to \sqrt{p}\exp(-i\theta_p)$, where "exp" represents the complex exponential function, $\exp(i\theta) = \cos(\theta) + i\sin(\theta) = e^{i\theta}$. With these character functions, it is easy to write down the corresponding L-function, of an elliptic curve E:

$$L(E, s) = \prod_p \frac{1}{(1 - \sqrt{p}e^{i\theta_p}p^{-s})(1 - \sqrt{p}e^{-i\theta_p}p^{-s})}$$

$$= \prod_p \frac{1}{1 - a_p p^{-s} + p \cdot p^{-2s}}.$$

Though this last expression looks formidable, it's actually not hard to compute the first few terms because one only needs to know the values a_p for any prime number p. As explained before, they can be computed by trial and error (if no better method presents itself). For example, the elliptic curve $y^2 + y = x^3 - x^2$ (otherwise known as curve 11a—the first curve listed in Figure 3) has the following Hasse-Weil L-function:

$$1 - \frac{2}{2^s} - \frac{1}{3^s} + \frac{2}{4^s} + \frac{1}{5^s} + \frac{2}{6^s} - \frac{2}{7^s} - \frac{2}{9^s} - \cdots .$$

It is *almost* true that the equidistribution of θ_p will follow from showing that $L(E, s)$ can be extended past the line $x = 1$ and that it never takes the value 0 on that line. The catch (as Serre realized in 1968) is that one actually needs to show these two facts not just for the Hasse-Weil L-function, but for an entire sequence of L-functions, in which the characters $p \to \sqrt{p} \exp(i\theta_p)$ and $p \to \sqrt{p} \exp(-i\theta_p)$ are replaced by their n-th powers. (These are the "L-functions of symmetric powers" that Taylor referred to in the quote above.) This additional complication arises because equidistribution for θ_p is much more complicated than equidistribution for primes in arithmetic progressions. In the example treated in the Box, "Dirichlet and the Distribution of Prime Numbers," involving the final digit of prime numbers, Dirichlet merely had to show that four events (the last digit is 1, the last digit is 3, the last digit is 7, or the last digit is 9) have equal probability. But in the Sato-Tate Conjecture, θ_p can take on infinitely many different values (any number between 0 and 2π) and we have to show they are all "equally likely." That is, θ_p is a *continuous* variable, not a discrete one. Therefore a proof that it is evenly distributed requires a more subtle analysis and, indeed, a more subtle definition of "evenly distributed."

Putting aside this extra complication for the moment, the question arises: How do you show that the Hasse-Weil L-functions can be extended to the left of the line $x = 1$, and that they have the desired property of being nonzero on that line? "You prove these things are modular, or automorphic," says Taylor. "That's the only way any subtle property of an L-function has ever been proved." And that brings us to the second of the three giant machines behind the proof of the Sato-Tate conjecture.

The theory of complex analytic functions provides a variety of techniques to extend an analytic function (defined by an infinite series) past the domain where its original definition converges. One way, for instance, is to come up with a closed-form expression for the series. A well-known example is the geometric series:

$$1 + z + z^2 + z^3 + \dots = \frac{1}{1 - z}.$$

The infinite sum on the left-hand side of the equation converges inside the unit disk $|z| < 1$, and at every point on the boundary except $z = 1$. It does not converge when $|z| > 1$, so for such values of z the equation above is meaningless. Nevertheless, the function on the right-hand side does make sense when $|z| > 1$, and so it provides a natural way of extending the function on the left-hand side. In fact, it is the *only* natural way to do so while preserving the function's differentiability properties.

Early analysts such as Leonhard Euler took this extension very literally and wrote formulas such as

$$1 + 2 + 2^2 + 2^3 + \ldots = -1$$

that do not quite pass today's standards of rigor.

What can we do if a closed form is not available for the series? Another option is to extend a function by symmetry properties. The simplest example is a theorem from complex analysis called the Schwarz reflection principle: if a complex analytic function maps a region whose boundary contains a line to another region whose boundary contains a line, then it can be extended in a natural way by "reflecting" the image. The figure on page 18 illustrates the Schwarz reflection principle for a harmonic function (the real part of a complex analytic function). If the blue part of the graph is given, then the orange part is the only way of extending it past the pink line.

The extension of L-functions from a half-plane to a whole plane borrows a little bit from both of these ideas. First, one relates the L-function to another kind of function, called a modular or automorphic function, which has what Mazur calls "hidden symmetry" properties. Then one can use those symmetry properties to extend the domain of definition of the L-function.

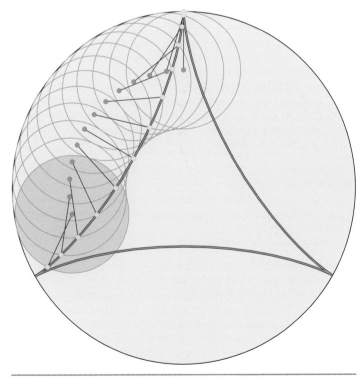

Figure 4. *A deltoid (the three-pointed curve formed by rolling a marked circle of radius 1/3 around the inside of a circle of radius 1) has a geometric symmetry that is far from obvious from the algebraic equation of the curve. Similarly, L-functions of elliptic curves have a geometric symmetry, called modularity, that is not at all obvious from their algebraic form. (Figure courtesy of Andrew D. Hwang.)*

An example will illustrate the general process. As noted before, the elliptic curve $y^2 + y = x^3 - x^2$ has an L-function that looks like this:

$$L(E, s) = 1 - \frac{2}{2^s} - \frac{1}{3^s} + \frac{2}{4^s} + \frac{1}{5^s} + \frac{2}{6^s} - \frac{2}{7^s} - \frac{2}{9^s} - \cdots .$$

A closely related function is the power series with the same coefficients:

$$f(q) = q - 2q^2 - q^3 + 2q^4 + q^5 + 2q^6 - 2q^7 - 2q^9 - \cdots .$$

This is called the *Mellin transform* of $L(E, s)$. The Mellin transform can be further modified by formally replacing the variable q by the variable $\exp(2\pi i z)$. The new function $f^*(z) = f(q)$ automatically inherits a nice symmetry property from the exponential function exp, namely it is periodic. That is, $f^*(z + 1) = f^*(z)$. But this particular function f^* also has a "hidden symmetry" property: It can be shown that $f^*(-1/11z) = z^2 f^*(z)$. This equation makes f^* a *modular form of level* 11 (because of the coefficient 11 in the "hidden symmetry" property) *and weight* 2 (because of the exponent 2 in the "hidden symmetry" property). (See Box, "Hidden Symmetries".)

Hidden Symmetries

There is, alas, no elementary way to see why the rather motley collection of coefficients in the L-function for an elliptic curve will produce a function with such an extraordinary symmetry. One can only say that algebraic expressions do not in general reveal the geometric symmetries that are hidden inside them. A simple example is the algebraic equation for a circle, $x^2 + y^2 = 1$, which reveals a few reflection symmetries of the circle (replace x by $-x$, or y by $-y$, or switch the variables x and y) but completely fails to reveal the infinitely many rotational symmetries of the circle. Or, for another example, consider the curve defined by the equation

$$y^4 + 2(x^2 + 12x + 9)y^2 + (x + 1)(x - 3)^3 = 0$$

(see Figure 4, page 27), called a deltoid. This curve, which is generated by rolling a circle of radius 1 around inside a circle of radius 3, has a beautiful threefold rotational symmetry. Who would have suspected this from the messy algebraic form of its equation?

Figure 5 (see next page) gives a hint of what the hidden symmetry of modular forms looks like. For each of the five elliptic curves listed in Figure 3, this figure shows a graph of the associated modular form over the complex unit disk. Because the modular form maps complex numbers to complex numbers, the full graph cannot be shown here; what is plotted instead is the complex argument of $f^*(z)$, or to put it a different way, the angle that the vector $f^*(z)$ makes with the x-axis. The red regions correspond to arguments between 0 and $\pi/6$, while the blue regions correspond to arguments between $-\pi/6$ and 0. All the shaded values have arguments that are near to a multiple of $\pi/6$. Notice that the behavior of $f^*(z)$ near the origin (six shaded lines cross; a red region and blue region share a common vertex; the red is situated clockwise from the blue) is

Cremona Number	Equation of Curve (Comments)	Graph of Curve (over real numbers)	Graph of modular form (over complex unit disk)
11a	$y^2 + y = x^3 - x^2 - 10x - 20$ First curve in Cremona's table. Has same modular form as $y^2 + y = x^3 - x^2$ (see text).	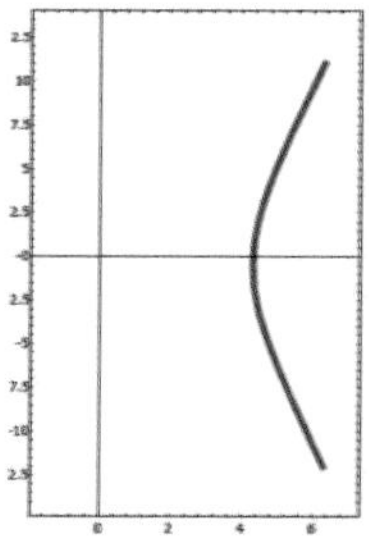	
36a	$y^2 = x^3 + 1$ Has complex multiplication. (Extra symmetry means that Sato-Tate conjecture does not apply.)	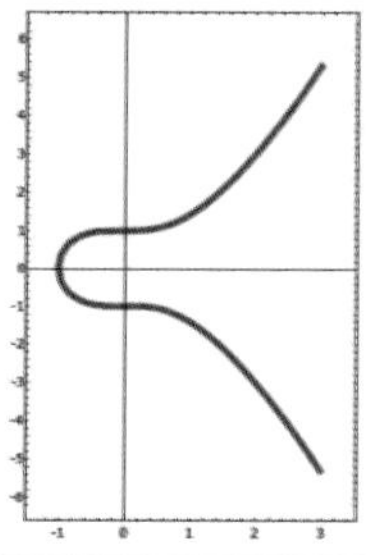	
96b	$y^2 = x^3 - x^2 - 2x$	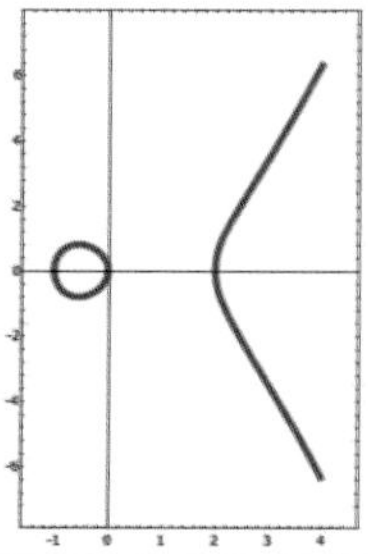	
198d	$y^2 + xy = x^3 - x^2 - 87x + 333$	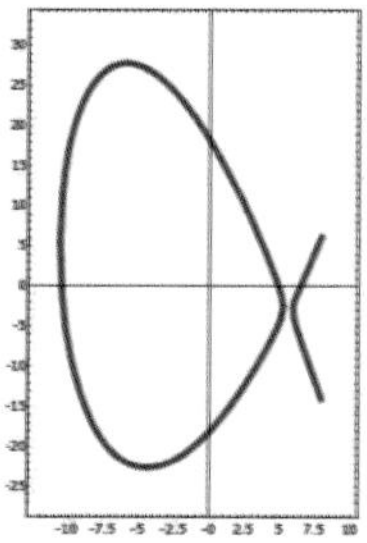	
8732a	$y^2 = x^3 - x + 9$ Reflection of Diophantus' elliptic curve (see text).	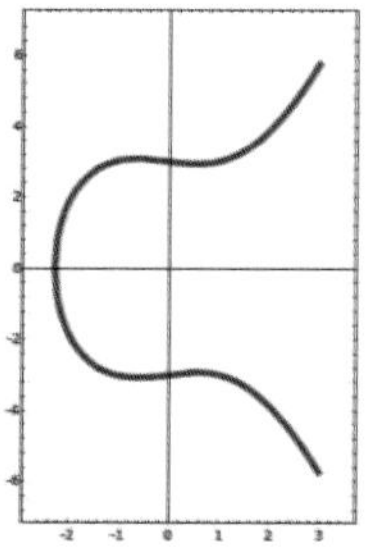	

Figure 5. *Five elliptic curves and the modular forms associated with them. (Figure courtesy of William Stein and Tomas J. Boothby, Dept. of Mathematics, University of Washington. The plots were done using the free open source program Sage (http://www.sagemath.org).)*

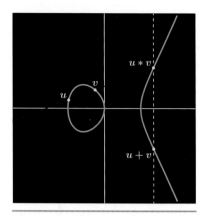

Figure 6. *The group law on elliptic curves. Note that the "sum" $u + v$ is different from the ordinary algebraic sum of u and v as complex numbers. (Figure courtesy of Andrew D. Hwang.)*

is repeated infinitely many times inside the unit disk. This infinite repetition is a direct consequence of the hidden symmetry of $f^*(z)$. (Note, incidentally, that the extra symmetry of curve 36a, which makes Sato-Tate inapplicable to it, also shows up in this figure. This modular form has a six-fold rotational symmetry that the other modular forms lack.)

To know that a certain algebraic expression possesses a hidden geometric symmetry is a rare and powerful gift. Modular forms of level N and height k can be completely tabulated by a nineteenth-century result called the Riemann-Roch theorem. So in this sense we have succeeded in reducing the L-function $L(E, s)$ to a "known" function, which is the first method of extending the domain of a complex function. Secondly, that known function has symmetry properties that themselves can be used to extend $L(E, s)$ beyond its original domain of definition. The symmetries can be thought of as a way of "cancelling out" the infinities that obscure the formal definition of $L(E, s)$, allowing us to discern the actual finite function inside.

However, this example leaves one question unanswered: Was it just a lucky accident that the particular elliptic curve $y^2 + y = x^3 - x^2$ was related to a modular form in this manner? The answer is no. The explanation requires us to explore the third of the three giant machines of contemporary number theory: Galois representations.

In 1993, Andrew Wiles astounded the mathematical world by claiming to have a proof of the most famous unsolved problem in mathematics, Fermat's Last Theorem (see *What's Happening in the Mathematical Sciences*, Volumes 2 and 3). His claim turned out to be not quite correct—there was a gap in his proof—but over the following year Wiles and Taylor were able to fill in the gap and complete the proof. But perhaps more importantly, the methods that Wiles introduced in 1993 have truly revolutionized the study of elliptic curves, and "enabled number theorists to win at error-term roulette" possible.

Fermat wrote his famous "Last Theorem" in the margin of his copy of Diophantus' *Arithmetica*, some time in the 1630s. (Because the *Arithmetica* was a book of problems, perhaps we can think of it as a "bonus problem"!) In modern terminology, Fermat asserted that there is no solution in nonzero integers a, b, c to any equation of the form $a^n + b^n = c^n$ (with n an integer greater than 2). Although Fermat claimed to have a proof, modern mathematicians are extremely skeptical of this claim. The proof eventually found by Wiles could not have been discovered in 1980, let alone 1630.

Wiles' approach to Fermat's problem was, paradoxically, to solve a *harder* problem and derive Fermat's Last Theorem as a special case. This strategy works amazingly often in mathematics—if you can find the right "harder problem," it often forces you to discover additional mathematical structure that was hidden in the original problem.

The "harder problem" chosen by Wiles was the Taniyama-Shimura Conjecture, which was formulated by a young Japanese mathematician, Yutaka Taniyama, in 1955. Taniyama tragically committed suicide in 1958, but the conjecture was refined (and, to some extent, corrected) by his colleague Goro Shimura. The conjecture simply states: *Every elliptic curve is modular.* In other words, the phenomenon we saw for the curve $y^2 + y =$

$x^3 - x$ and for the other curves in Figure 6 was no fluke. The L-function of *any* elliptic curve coincides with the L-function for some modular form. (Technically, the elliptic curve may need to be replaced first by an "equivalent" curve, related to the first by a mapping that number theorists call an isogeny. This is only a minor complication.)

It's not obvious how the Taniyama-Shimura Conjecture relates to Fermat's Last Theorem because the equation $x^n + y^n = z^n$ does *not* reduce to an elliptic curve except in the cases $n = 3$ and 4. However, in the 1980s Gerhard Frey and Kenneth Ribet showed that if a, b, and c were solutions to Fermat's equation, then the elliptic curve $y^2 = x(x - a^n)(x + b^n)$ would have some very strange properties. In particular, such a curve would not be modular, and hence would violate the Taniyama-Shimura Conjecture. If someone could prove the Taniyama-Shimura Conjecture, it would then follow that no such numbers a, b, c could exist.

Most number theorists thought that the Taniyama-Shimura Conjecture would remain unproved for many years. But Wiles did not agree. By 1993, he managed to prove enough of the conjecture to apply it to Frey's elliptic curve, and solve Fermat's Last Theorem. The full proof of the Taniyama-Shimura conjecture was completed in 1999 by Taylor, Christoph Breuil, Brian Conrad, and Fred Diamond.

The machinery that Wiles perfected, called Galois representations, uses a property of elliptic curves that has not yet been mentioned. Every elliptic curve is actually a group—which means that if you are given two rational points on an elliptic curve, A and B, you can discover a third point, A "+" B. This operation, "+", satisfies all the normal properties of addition, but is defined in a vastly different way, based on geometry. To find A "+" B, you draw a line (or "chord") through A and B. This line is guaranteed to intersect the curve in a third point, A∗B. Then you draw a vertical line through A∗B, which is guaranteed to intersect the elliptic curve in another point, A "+" B. (See Figure 6.) If A and B are the same, the "line through A and B" is actually the tangent to the elliptic curve through A.

Amazingly, Diophantus himself, 1750 years ago, had discovered the chord-and-tangent construction, although he surely didn't think of it that way. To solve his equation, $y(6-y) = x^3 - x$, he started with the "obvious" solution A = (x, y) = $(-1, 0)$. He then performed a calculation that is mathematically equivalent to computing A∗A = $(17/9, 26/27)$. (See Figure 7.) If he had wanted to, Diophantus could have repeated the procedure to generate many more rational solutions.

This group law begs to be used to solve the Taniyama-Shimura Conjecture. After all, shouldn't one form of geometric symmetry beget another?

Here is the ingenious way that Wiles did it. We have already seen one way of "finitizing" an elliptic curve—by reducing it modulo a prime, p. The group law gives us a second way of finitizing it—to look only at the points A that come back to themselves after a finite number of iterations of the chord-and-tangent law. In other words, consider the points $E[n]$ for which

$$A`` + ``A`` + ``\ldots`` + ``A = A,$$

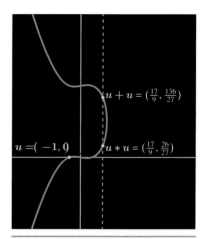

Figure 7. *The group law in action. One point on Diophantus' curve, $u = (-1, 0)$ is easily found by inspection. Diophantus found a second rational solution, $u * u = (17/9, 26/27)$, by a method that was algebraically equivalent to the chord-and-tangent construction illustrated here. The sum $u + u$ is yet another rational solution. (Figure courtesy of Andrew D. Hwang.)*

This is only a very sketchy outline of a tremendously complex and innovative argument. However, yesterday's novelty becomes today's common knowledge...

or "n times A" equals "O". (All these operations are in quotation marks because they are not the normal operations of addition and multiplication, and the point "O" on the elliptic curve is definitely not zero—it's actually the "point at infinity"!) Wiles hit on the idea of finitizing in both ways at once—reducing modulo p and looking only at the points $E[l]$, where l is a prime number different from p. From the algebraic point of view, the resulting group is surprisingly simple: it looks like a finite (l-by-l) piece of a 2-dimensional lattice.

The two functions, $p \to \sqrt{p}\exp(i\theta_p)$ and $p \to \sqrt{p}\exp(-i\theta_p)$, which were mentioned before as being "like" Dirichlet characters, turn out to be eigenvalues of a 2×2 matrix that acts on the 2-dimensional lattice $E[l]$. The trace of this matrix (which is also called a *character of a Galois representation*) is just the function $p \to a_p$ whose distribution is at issue in the Sato-Tate Conjecture. In this way a link is established between the three "giant machines" because the function $p \to a_p$ has now appeared in three different contexts:

• As the coefficients of the L-function of an elliptic curve E;
• As the coefficients of the L-function of a modular form f^*;
• As the character of a Galois representation (or the trace of 2×2 matrices acting on the lattice $E[l]$ reduced modulo p).

Of the three machines, the third is ideal for proving the Taniyama-Shimura Conjecture because it is here that the "contagiousness" of modularity becomes apparent. First, Wiles showed the modularity of $E[l]$ modulo p; then he discovered a method, called "modularity lifting," to prove that $E[l^\nu]$ modulo p is modular, for all powers l^ν of the prime l; and finally, he showed that the modularity contagion spreads to different primes p, and then back to different primes l. After it has been proved that $E[l^\nu]$ modulo p is modular for all primes l and p, there is nothing left to prove; E is modular.

This is only a very sketchy outline of a tremendously complex and innovative argument. However, yesterday's novelty becomes today's common knowledge, and in retrospect, Wiles had it easy in two ways. First, he actually needed to use only two primes l and p: namely, 3 and 5. Second, he only had to worry about 2×2 matrices. This was important because the group of (invertible) 2×2 matrices modulo 3, $GL_2(F_3)$ is very small, and in particular it possesses a property called *solvability* that larger matrix groups would not have had. "It happened that for this group, some results of Langlands and Tunnell applied. It was a sort of coincidence, and that coincidence no longer works for GL_n [i.e., $n \times n$ matrices]," says Taylor. "For us, working on the Sato-Tate Conjecture, the first step was completely missing. For the second step, though it's hard, we copied what Andrew did. Andrew's argument is a sort of inductive argument, but he was lucky that the first step was there in the case he considered."

To prove the Sato-Tate Conjecture, Taylor and his colleagues needed a version of the Taniyama-Shimura Conjecture that would apply to characters of $n \times n$ matrices. The reason goes all the way back to Serre's comment about symmetric powers. Remember that it is not good enough to prove that the characters $\exp(i\theta_p)$ produce a modular L-function; you also have to do the same thing for all of their powers, $\exp(in\theta_p)$. Other mathematicians had managed to prove this for $n = 2, 3,$

and 4, but by a very laborious method. In essence, the other mathematicians had been taking the output of the three giant machines and trying to re-shape it to fit the Sato-Tate Conjecture. Instead, Taylor realized, you had to go under the hood of the machines—specifically, the third machine—and tinker with it to make it work for $n \times n$ matrices. Who could be better qualified to do that than the person who helped Wiles build the machine in the first place?

Even so, Taylor needed some outside help. In particular, the problem of finding a "starting point" for the modularity lifting argument led him to an unexpected source. "We used a family of Calabi-Yau varieties that we discovered in the physics literature," Taylor says. Nick Shepherd-Barron, in particular, pointed Taylor and Harris in the right direction. Armed with the physicists' calculations, and with the new concept of "potential modularity" that Taylor had invented, the team of four mathematicians (Taylor, Shepherd-Barron, Harris, and Clozel) finally closed in on the proof of the Sato-Tate Conjecture in 2006. Like the Taniyama-Shimura Conjecture before it, Sato-Tate has given up very grudgingly: The proof took ten years from the time that Taylor and Harris began working on it. Perhaps the biggest sticking point was to show that "potential modularity" is contagious in the same way that modularity is. Even now, there are still a few elliptic curves for which the proof doesn't hold. But for a satisfyingly large class of elliptic curves, it does. "Before this was proven, there wasn't a single elliptic curve for which we knew the Sato-Tate Conjecture!" says Katz.

Now that the Sato-Tate Conjecture has become a theorem (at least for most elliptic curves), there are still plenty of intriguing questions left. Mazur cites this example: How many ways are there to represent a prime number p as a sum of 24 squares? The "expected answer" is amazingly simple: $\frac{16}{691}(p^{11} + 1)$. The "random errors" are known to be less than $\frac{66304}{691}\sqrt{p^{11}}$. The errors seem, in computer experiments, to follow a Sato-Tate distribution. Yet there is no elliptic curve involved, and no apparent way to modify Taylor's proof of the Sato-Tate conjecture to apply to this problem.

Mazur also points out that, having found the mean and the shape of the error distribution, number theorists would now like to know how rapidly the observed frequencies for a_p (or θ_p) actually approach the Sato-Tate shape. Number theorists have generated lots of computer data and some specific conjectures, but Mazur says, "We don't even have a scenario for a proof yet."

"We're still, in a sense, scratching the surface," says Taylor. "These ideas should go much further. When we can pick off something nice like Sato-Tate, it's proof that we are making progress."

John Tate. *(Photo courtesy of the American Mathematical Society.)*

Pizza Toss. *World pizza champion Tony Gemigniani demonstrates his pizza-tossing form. In a pizza toss, the angular momentum vector is very close to the normal vector. Thus the pizza never turns over, and always lands "heads". (Photo courtesy of Tony F. Gemigniani, President, World Pizza Champions, Inc.)*

The Fifty-one Percent Solution

Persi Diaconis and Susan Holmes.

FOR CENTURIES, COINS HAVE BEEN AN ICON of randomness. Who hasn't flipped a coin to decide between two equally appealing alternatives—which restaurant to go to, which road to take? Especially when the choice doesn't matter too much, tossing a coin beats thinking.

But sometimes, even important choices are left to the caprice of a coin. In 1845, two settlers in the Oregon Territory, Asa Lovejoy and Francis Pettygrove, founded a new town on the banks of the Willamette River. Lovejoy wanted to name it after his birthplace, Boston, but Pettygrove preferred to name it after his own hometown—Portland, Maine. A coin toss seemed like the fairest way to settle the dispute. The penny landed in Pettygrove's favor, and that is why the largest city in Oregon is now called Portland, instead of Boston.

The implicit assumption behind coin flips has always been that that they are fair—in other words, that heads and tails have an equal chance of occurring. However, a team of three mathematicians has now proved that this assumption is incorrect. Any coin that is tossed vigorously and high, and caught in midair (rather than bouncing on the ground) has about a 51 percent chance of landing with the same face up that it started with. Thus, if you catch a glimpse of the coin before it is

Figure 1. *A stroboscopic photo of a coin toss over a period of one second. Between the toss and the first landing, the coin made two full revolutions (or four half-revolutions), and thus the upward face was alternately heads-tails-heads-tails-heads. Thus, it landed in the same orientation that it started, a result that, according to new research, happens about 51 percent of the time. (Photo courtesy of Andrew Davidhazy, Rochester Institute of Technology.)*

tossed and see heads up, you should call heads; on the other hand, if you see the tails side up, you should call tails. This will give you a 51 percent chance of predicting the outcome correctly.

Persi Diaconis and Susan Holmes of Stanford University, together with Richard Montgomery of the University of California at Santa Cruz, published their discovery in *SIAM Review* in 2007, although it was originally announced in 2004. According to Diaconis, who has given public lectures on the topic numerous times, the result is a difficult one for most non-mathematicians to grasp. The coin's bias has nothing to do with whether it is weighted more on one side or the other, nor with any asymmetry in its shape. In fact, their analysis assumes that the coin is perfectly symmetrical. The coin's bias lies not in its shape, but in its motion—the dynamics of a rigid, rotating object.

From a physicist's point of view, a coin's motion is completely deterministic. If you know the initial orientation, velocity, and angular velocity of the coin, you can predict its future flight perfectly. Diaconis, in fact, has built a coin-tossing machine that illustrates this fact (see Figure 2). The machine tosses heads with spooky consistency. You press a lever, it launches a coin into a collecting cup, and the coin will come up heads every single time. There's no chance involved.

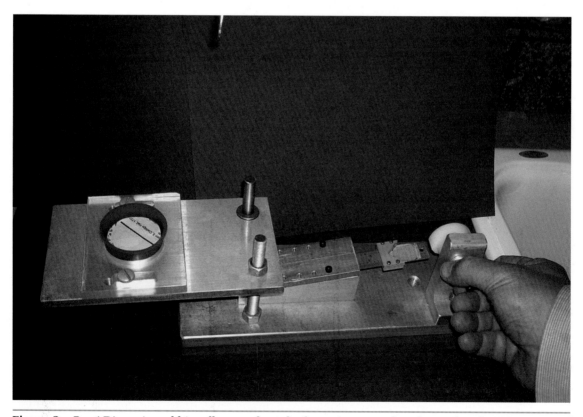

Figure 2. *Persi Diaconis and his colleagues have built a coin tosser that throws heads 100 percent of the time. A coin's flight is perfectly deterministic—it is only our lack of machine-like motor control that makes it appear random. (Photo courtesy of Susan Holmes.)*

If a machine can produce heads 100 percent of the time, why do humans have such blind faith in the coin's randomness? In a sense, it is not the coin's randomness that is at issue, but our own clumsiness. To produce heads all the time, you need extremely precise control over the coin's initial conditions, such as the strength of the toss and the rotation rate imparted to the coin. Humans do not normally have such fine motor control. Even so, there are some ways that humans can, either intentionally or unintentionally, bias their coin tosses.

First, if you toss the coin in the same way that a pizza maker tosses his pizza dough (see **Pizza Toss**, page 34)—setting it in motion around its normal axis instead of its diameter—then the coin will never flip over, and you will get heads 100 percent of the time (assuming the coin starts with heads up). Magicians have perfected a trick based on the "pizza toss." They realized that they could add a little wobble, but not too much. The casual viewer cannot tell the difference between the wobbly "pizza toss" and a real coin flip, and the magician will still get heads 100 percent of the time.

A second form of bias occurs if you give the coin a very wimpy toss, so that it rotates only a half-turn before landing. The coin will always land with the opposite face pointing up, and so a sequence of "wimpy tosses" will go heads-tails-heads-tails forever. Even though the percentage of heads is exactly 50 percent, these tosses are far from random.

Diaconis once received a class project, called "The Search for Randomness," that fell into this second trap. A teacher had asked his students to flip a coin 300 times each, thus generating a table of 10,000 supposedly random coin flips, which he sent to Diaconis. Alas, Diaconis says, "The results were very patterned. The reason was that the students got bored." (Wouldn't you, if you were flipping a coin for a full class period?) The more bored the students got, the wimpier their flips were, and the more frequently the telltale pattern heads-tails-heads-tails started to show up.

Both of these exceptions turn out to be very important for the mathematics of coin-flipping because they are extreme cases. The first exception shows that it makes a difference what axis the coin rotates about. The second shows the importance of making sure the coin spins a reasonably large number of times before it is caught. By trickery or by simple disinterest, a human can easily manage a biased coin flip. But the question Diaconis wanted to know was: What happens if an ordinary human honestly tries to achieve a random and unbiased flip? Can he do it?

Clearly, the flip should rotate the coin about its diameter, and it should be vigorous enough to ensure a fairly large number of revolutions. Under those two conditions, Joe Keller, an applied mathematician at Stanford, showed in 1986 that the probability of heads does indeed approach 50 percent. However, this was a somewhat idealized result because Keller assumed that the number of revolutions approaches infinity. Diaconis' student Eduardo Engel followed up on Keller's work by studying what would happen under more realistic conditions—a coin that rotates between 36 and 40 times per second, for about half a second (to be precise, between .44 and .56 seconds). Engel, who is now at Yale University, showed that the

If a machine can produce heads 100 percent of the time, why do humans have such blind faith in the coin's randomness? In a sense, it is not the coin's randomness that is at issue, but our own clumsiness.

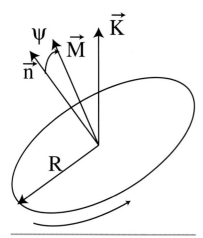

Figure 3. *The three relevant vectors for determining how a coin will land are the normal vector (n), the angular momentum vector M, and the upward vector K. The vector M remains stationary, and n precesses around it in a cone. Thus, the angle between M and n remains constant (ψ). The coin will turn over only if the angle between n and K exceeds 90 degrees at some time. In this sketch, because ψ is small, the coin will never turn over—this is a "pizza toss". (Figure courtesy of Susan Holmes.)*

probability of heads under these circumstances (assuming the coin starts with heads up) lies between 0.444 and 0.556. Engel's result was already quite sophisticated, requiring careful estimates of integrals over the space of allowable initial conditions. Neither he nor Keller discovered any indication that coin tosses are biased.

However, there is a second simplifying assumption in both Keller's and Engel's work. They assumed that the flip sets the coin spinning *exactly* around its diameter (i.e., the axis of rotation lies exactly in the plane of the coin). Even if the coin-flipper is trying to give the coin an honest flip, it is unrealistic to expect such perfection. What happens if the coin rotates about an axis that is neither perpendicular to the coin as in the "pizza toss," nor exactly in the plane of the coin as in the "Keller toss," but instead points off in some oblique direction? That is exactly when the problem gets interesting.

In 2003, Diaconis visited Montgomery in Santa Cruz, and saw on the wall of his office a poster depicting how a falling cat rotates itself to land on its feet. Suddenly he knew he had found the right person to ask about the rotation of a falling coin.

Coins are actually simpler than cats because they are rigid objects, and also because they have circular symmetry. Cats, according to Montgomery, are best viewed as a system of two rigid objects (the cat's front and rear) with a flexible link between them. The flexible link—the cat's muscles—gives it control over its landing orientation.

The coin, on the other hand, has no such control. Its motion is completely determined from the moment it is tossed to the moment it is caught. The motion of a "free rigid object," as physicists call it, has been understood ever since Leonhard Euler in the eighteenth century. Although, the description of the motion is quite simple, it is perhaps not as familiar as it should be.

From the point of view of a physicist, the key parameter for describing the coin's motion is its angular momentum—or what we have loosely referred to as the "axis of rotation" above. To the untrained observer, the coin's behavior looks like a complicated combination of spinning, tumbling, and wobbling, but to a physicist, all of these kinds of motion can be summed up in one vector, the angular momentum. Moreover, the angular momentum vector remains unchanged for the entire time that the coin is in the air.

What matters most for the outcome of the coin toss, though, is not the orientation of the angular momentum, but the orientation of the coin's normal vector. (This is the unit vector pointing perpendicularly to the coin, in the "heads" direction, as shown in Figure 3.) If the coin is caught with the normal vector pointing anywhere in the upper hemisphere, it will be interpreted as a "heads" flip. If the normal vector points towards the lower hemisphere, the flip will be recorded as "tails."

The normal vector and the angular momentum vector are different, but they are related in an exceptionally simple way. The angle between them, denoted ψ, remains constant. This means that the normal vector *precesses* around the fixed angular momentum vector, forming a cone with an opening angle of ψ radians. Alternatively, when the trajectory of the unit normal vector is plotted on a sphere, the result is a circle of radius ψ

whose center is the angular momentum vector (normalized to have length one).

From this point of view, the "pizza flip" and the "Keller flip" are particularly simple cases. In the "pizza flip," the angular momentum vector and the normal vector coincide. The angle between them is 0. Because the angular momentum is constant, and the normal vector moves in a "circle of radius 0" around it, the normal vector is also constant. If the coin starts out heads (i.e., with normal vector $(0, 0, 1)$) then the normal vector will continue to point in the same direction, and the coin will always land heads.

In the "Keller flip," on the other hand, the angular momentum vector lies in the plane of the coin, making an angle of $90°$ or $\pi/2$ radians with the normal vector. Hence the normal vector precesses around a circle of radius $\pi/2$—in other words, a great circle on the unit sphere. If we assume that it starts pointing in the direction $(1, 0, 0)$, then it will rotate at a constant rate from the north pole to the south pole and back. Half of the time the normal vector will be in the upper hemisphere, and half of the time it will be in the lower hemisphere. In a rough sense, this explains why the coin has a 50 percent chance of landing heads and a 50 percent chance of landing tails (as Keller proved).

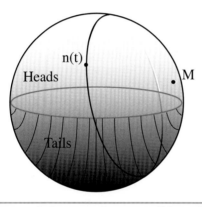

Figure 4. *In general, the normal vector will describe a circle centered at M while the coin is in the air. If the coin starts perfectly horizontal (so that n is perfectly vertical), then n will inevitably spend more time in the upper hemisphere than the lower hemisphere. If the coin is caught at a random time, it will therefore have a greater than 50 percent probability of being caught with n in the upper hemisphere. (Figure courtesy of Susan Holmes.)*

In a real flip, the normal vector will precess in a circle of radius ψ radians, a "small circle" rather than a great circle (See Figure 4). If the coin is tossed from a heads-up position, then this small circle passes through the point $(1, 0, 0)$. If ψ is less than $\pi/4$ radians, then in fact the entire small circle lies in the upper hemisphere. This means that the coin never actually turns over; it will land heads with 100% probability. This is why magicians do not have to execute a perfect "pizza toss" in order to ensure the coin lands heads; there is in fact considerable room for error. A coin tossed with ψ close to $\pi/4$ will precess so vigorously that the eye is easily deceived into thinking that it is tumbling, and yet it will always land heads.

As beautiful as it is, this analysis did not allow the mathematicians to make a precise estimate of the amount of bias in a human-tossed coin ... To obtain a quantitative estimate, therefore, the mathematicians had to study empirically how humans actually flip coins. And that was not nearly as easy as it sounds.

If ψ lies between $\pi/4$ and $\pi/2$, then the normal vector will sometimes enter the lower hemisphere. Thus there will be a nonzero chance of the coin landing tails. However, in all cases the normal vector will spend more time in the upper hemisphere than in the lower hemisphere, and therefore the probability of heads will always exceed 50 percent. Although this is a somewhat informal argument, Diaconis, Holmes and Montgomery formalized it in the same way that Keller did, by taking a limit as the time of flight goes to infinity. The probability of heads (assuming that the coin starts with heads up) is

$$p(\psi, \phi) = \frac{1}{2} + \frac{1}{\pi} \sin^{-1}(\cot \psi \cot \phi),$$

where ϕ is the (fixed) angle that the angular momentum vector makes with the vertical. If the inverse sine is undefined, as is the case when ψ is small, then the probability of heads is simply 1.

As beautiful as it is, this analysis did not allow the mathematicians to make a precise estimate of the amount of bias in a human-tossed coin. That depends on how close ψ is to $\pi/2$ (i.e., how closely the coin flipper can approximate a perfect "Keller flip"). To obtain a quantitative estimate, therefore, the mathematicians had to study empirically how humans actually flip coins. And that was not nearly as easy as it sounds.

"When Joe [Keller] had written his paper, I wanted an answer to the question of how many times the coin turns," Diaconis says. "At that time, I found out that there was only one slow-motion camera at Stanford, which was owned by the football team. In order to use it, I would have to pay the operator something like $2800 for a two-hour session. I wanted to know the answer, but I didn't want to know it that badly!"

A decade and a half later, after they had talked with Montgomery, Diaconis and Holmes again tried to videotape real coin flips. But an ordinary video camera is far too slow: it shoots 60 frames per second. Because a typical coin makes at least 20 full revolutions per second, a frame-by-frame data set is far too coarse to tell what it is actually doing. "You'd like to have up to 800 frames per second," Diaconis says. They tried to overcome the problem by using eight videocameras at once, but the logistics proved to be too difficult. "I was in despair," Diaconis says. But this time, he found out that Abbas El Gamal, of Stanford's electrical engineering department, had an ultra-slow-motion camera. Unlike the football team, El Gamal was happy to have his camera used for coin-flipping research. Problem solved!

In reality, that was only the beginning. From the physical point of view, it wasn't so easy to flip a coin in such a way that the camera could record its flight successfully. The camera was stationary, so the flip had to pass directly in front of the camera lens. And the camera would only record for about a quarter of a second, so the flip had to be synchronized very closely to the start of filming. Out of 50 attempts, only 27 gave useful results.

Then there were mathematical challenges, some with very ingenious solutions. First was the problem of simply finding the coin in each of the images. With 27 videos of 100 frames each, it would be impractical to draw the outline of the coin on each frame by hand. Moreover, the human eye and hand are not objective enough. Holmes tried out several different statistical learning algorithms to find one that would be able to pick out the coin pixels from the background pixels; she then used standard statistical techniques to find the ellipse that most closely matched the edge of the coin. Finally, she could use this ellipse to determine which way the normal vector was pointing at each instant.

At this point, a surprise awaited her. When plotted on a sphere, the normal vectors did not lie along a circle (the circle of precession), as theory said they were supposed to. Instead, they just made a formless blob. (See Figure 5.) "It was devastating the first time we saw the data," Holmes recalls.

Finally, Holmes figured out what the problem was. The two-dimensional photographic image of the coin does not completely determine its orientation in space. For any given ellipse, there are four possible orientations of the coin that would produce that ellipse: the nearer edge of the coin could be either above or below the farther side, and the top face could either be heads or tails. If you know the coin's history, you can tell which orientation is correct because the coin won't suddenly jump from one orientation to another. But teaching the computer to make this determination automatically took a little bit of work. Finally Holmes was able to unscramble the data enough to get a satisfactory approximation of a circle. Once the circle had been determined, it was easy to find the angular momentum vector (the center of the circle) and the parameter ψ (the angular radius of the circle).

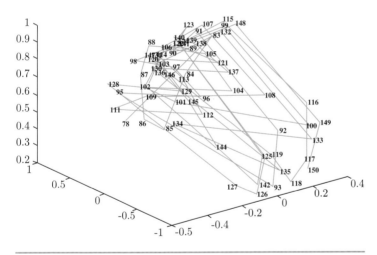

Figure 5. *At first, the photographic data seemed to show the normal vectors traveling in a strange, jerky pattern—not in a circle as Figure 4 would predict. However, the researchers eventually figured out that the problem lay in the ambiguity of a 2-dimensional photographic image. The normal to the coin could be pointing in any of four directions in 3-space, and their automated image analysis program was picking the wrong ones. (Figure courtesy of Susan Holmes.)*

As it turned out, for some of the tosses there was an independent way to check the computation of ψ. It's an intriguing, and not obvious, physical fact that each time the coin's normal vector precesses exactly one time about the angular momentum vector, the coin also rotates a fixed amount (in a coin-centered coordinate frame) about the normal vector. Figure 6 shows a nice example: the coin was imaged three times facing in the same direction. (The images were separated by 20 frames, and therefore by 1/30 of a second). Between each pair of frames, the letter "T" in the center of the coin has rotated through an equal angle. This angle of rotation, ΔA, is related to ψ by the following equation:

$$\Delta A = -(1 - I_1/I_3)2\pi \cos(\psi),$$

where I_1 and I_3 represent the moments of inertia of the coin about its diameter and normal, respectively. For a very thin coin, the ratio I_1/I_3 is $\frac{1}{2}$, but for a U.S. half dollar, the ratio is about 0.513.

The effect is well known to quantum physicists, who call it the *Berry phase* after Michael Berry, who (re-)discovered it in 1984. It can arise whenever the wave parameters of an oscillating system are changed slowly and then returned to their original values. The system will appear to have returned to its original state, but it will have accumulated a phase difference that can be detected (in the quantum physics applications) by using an interferometer. Curiously, Montgomery had just written a book with a chapter devoted to Berry phases—another way in which he turned out to be just the right man for the project.

The Berry phase gave Diaconis, Holmes, and Montgomery an independent way of computing the angle ψ, and thereby validating that their image analysis algorithm was working correctly. With their computation of ψ and the angular momentum vector, they could use Diaconis' formula for $p(\psi, \phi)$ to compute the probability of heads on each of the 27 flips that they had recorded. (Note that the actual outcome of the flips was immaterial—it was the probability that they wanted.) The average probability was 0.508, which they rounded up to 0.51, and this was the basis for their claim that real coins have a 51 percent chance of landing with the same side up that they began with.

The result comes with a number of caveats, none of which alter the basic conclusion that coin tosses are biased. First, it is exceedingly important to catch the coin in midair, and not let it hit the ground. Once it hits the ground, other factors—such as the shape or weight distribution of the coin—start playing a role. Magicians have learned how to shave the edge of a coin so that if you spin it on a table (rather than tossing it in the air) the coin will always come up heads. Although letting the coin bounce may be acceptable for a perfectly symmetric coin, it is too easy to tamper with the coin and change the probabilities.

Also, the 51 percent probability is only an ideal estimate, when the coin is allowed to fly for a long time. Real coin flips tend to last only a half second or so. The finite-time-of-flight effect means that the actual probability of heads falls in a range of values (just as Engel's analysis of Keller flips gave the probability a range of values). However, the range will be centered on 51 percent, not on 50 percent.

Third—and this is the point that worries Diaconis the most—it's still unclear just how representative their 27 tosses were of what happens in a real coin toss. Because it was so hard to synchronize the tosses with the high-tech camera, the tosses were probably performed more carefully and perhaps less vigorously than real coin tosses. Thus it is quite possible that the real bias in favor of the starting position is more than 51 percent, and indeed quite a bit more. This became apparent in a "low-tech" experiment that Diaconis performed, attaching a ribbon to the coin so that he could count how many times it flipped over. (The number of flips is the same as the number of twists in the ribbon when the coin is caught.) In 4 out of 100 tosses, the coin never flipped at all—Diaconis had unintentionally performed a "pizza flip." In 3 out of 100 tosses, the coin flipped only once. Thus a significant number of real-world coin flips may be executed unintentionally in a way that makes them far from random.

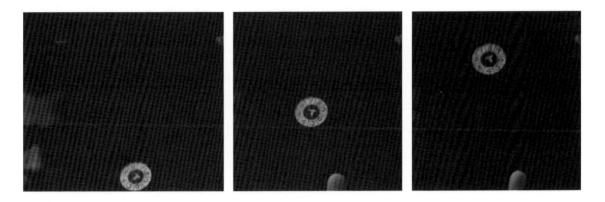

FRAME 48   FRAME 68   FRAME 88

Figure 6. *In this trial, the tosser happened to flip the coin at very close to 30 revolutions per second. Thus, with a high-speed camera that photographed the coin 600 times per second, the coin was nearly in the same orientation every 20 frames. Notice that while the coin has made a complete revolution, it has precessed around its normal axis by less than a full revolution. In fact, the amount of precession, called the Berry phase, provides an independent way to measure the angle ψ mentioned in Figure 3. (Figure courtesy of Susan Holmes.)*

Montgomery's wife Judith, a school math teacher, has suggested holding a "great California flip-off" to see if real-world coin flips actually do have the predicted amount of bias. It will take a lot of flips, though. To verify experimentally that the probability of heads is *not* 50 percent but 51 percent, one would need a random sample of about a quarter-million flips. Given the difficulty of even obtaining a sample of 10,000 flips (remember the bored math students?), Diaconis is not optimistic about the chances of pulling off such a massive experiment. "The idea of performing quality control on that data puts me off," he says.

For all three researchers, the coin-flipping work has led in unexpected directions, some of them much more serious than the original problem. To Montgomery, it's a great teaching example for a course in differential geometry. The rotations of the coin correspond to elements of the Lie group SO(3), which can be associated to points on a hypersphere (that is, a three-dimensional sphere in four-dimensional space). Different shapes or weightings of the coin correspond to different ways of measuring distance on the hypersphere, and the complicated dance of the coin always represents the shortest path through this three-dimensional space.

For Diaconis, a sequence of coin flips is an analogue for a much more complicated process—namely, the folding of a protein. The coin exhibits a very simple kind of dependence between its successive states—namely, it has a 51 percent chance of staying in the same state it was in (heads or tails), and a 49 percent chance that it will switch to the opposite state. Yet determining this dependence from first principles was not at all easy. A protein molecule has many more configurations or states; it's not just a simple dichotomy of "heads" and "tails." The challenge is to simplify the description of the molecule so that the number of states is manageable, the states are still physically meaningful, and one can compute the probability of moving from one state to another. "The coins were a direct motivation for our work on protein folding," Diaconis says.

Finally, for Holmes, the "baby problem" of automatically detecting a coin in a photograph has led to a new interest in image analysis. She is currently working with a software package called Gemident, to train computers to recognize cancer and immune cells in an image of a lymph node (Figure 7). "In the beginning, everyone counted these cells by hand, and it wasn't an objective method," she says. The automated detection algorithm makes it possible to do quantitative statistical analysis of the images—for example, determining where the cancer cells lie in relation to the immune cells (such as T-cells and B-cells).

"The value [of the coin-flipping project] for me was that it showed me it was doable, that you can work with immense computer files like these," Holmes says. "Now I'm able to say to the biologists, 'You shouldn't do this by hand, because I can teach the computer to do it.'"

Perhaps they will follow in the footsteps of Richard Feynman, the physicist who first worked out the theory of quantum electrodynamics—and who said that his theory was inspired by watching a dinner plate that was tossed into the air. In *Surely You're Joking, Mr. Feynman*, the Nobel laureate recalls, "The [Feynman] diagrams and the whole business that I got the Nobel Prize for came from that piddling around with the wobbling plate." Inspiration for big things can indeed come from humble sources, even the common flip of a coin or a dinner plate.

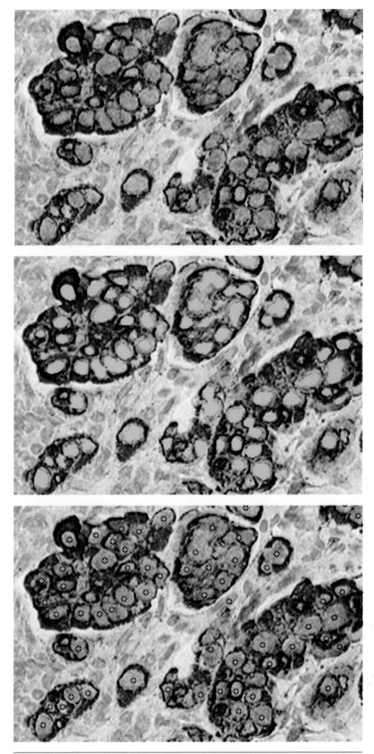

Figure 7. *Since the group's coin experiments, Holmes has worked on other projects involving automated image analysis. Here, she has "taught" a computer to recognize cancerous cells (indicated by green dots) in a microscopic image of a lymph node. (Figure courtesy of Susan Holmes.)*

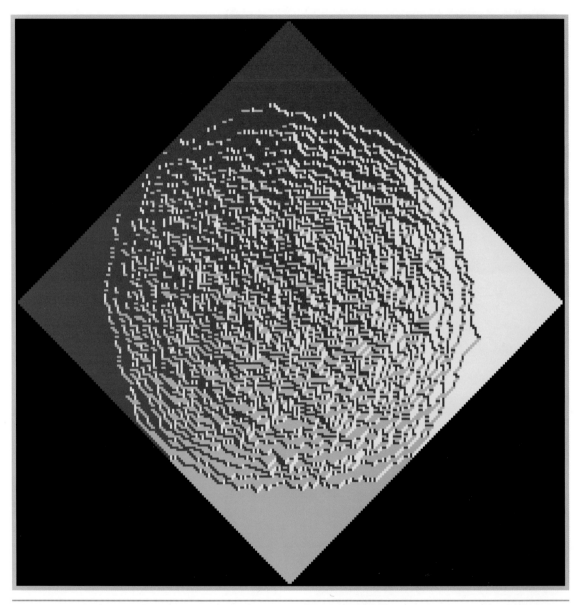

Aztec Diamond. *A random domino tiling of an Aztec diamond of order 256. The regions at the corners are "frozen," with all the tiles in the same orientation. The central region is "liquid," and the boundary between the two looks roughly circular. The Arctic Circle Theorem states that the boundary approaches a circle as the order of the diamond approaches infinity. (Figure courtesy of Cristopher Moore.)*

Dominos, Anyone?

AN OLD MATHEMATICAL CHESTNUT runs as follows: Suppose you remove two diagonally opposite squares of a chessboard. Can you then tile the remaining part of the board with dominos? In other words, can you put down 31 dominos so that they cover all of the remaining squares, without overlapping or leaving any square vacant?

Like all good puzzles, this one seems hard at first, but it has an easy solution if you think about it the right way. First, you have to realize that each domino must cover one white square and one black square. But the two removed squares have the same color, and therefore there is an imbalance of colors remaining: either 30 white squares and 32 black, or vice versa (depending on which corners you removed). So after you've laid down the first 30 dominos, you will either have two black squares left or two white squares, and therefore they will not be coverable by a single domino. Thus, the tiling is impossible. (See Figure 1.)

Remarkably, this simple puzzle has turned into a prototype for a burgeoning area of mathematics in recent years—the study of random tilings or (after a small change in perspective) random surfaces. Random tilings by dominos, as well as other shapes, have turned into a beautiful model for phase transitions in physics. Certain random tiling models are simple enough to be solved explicitly, allowing precise deterministic predictions. Yet at the same time they are complicated enough to display "solid," "liquid," and "gaseous" states—much to the surprise of the experts who started working on them in the 1990s. By contrast, more realistic 3-dimensional models and even some other 2-dimensional models of phase transitions do not (yet) have closed-form solutions.

"That's why everyone likes them so much," says Scott Sheffield, a mathematical physicist at New York University. "There are a lot of things one would expect to be true in general for random surfaces, but there's no hope of proving them. Random tilings have a lot of behavior you would expect the other models to have, and they seem to capture the important features of what you want a random surface to be."

After solving the brain-teaser about the chessboard with the corners removed, the next and more challenging problem is to count *how many* domino tilings there are for a given feasible region. For example, we know it is feasible to tile an entire chessboard with 32 dominos; how many different ways are there to do it?

This problem can be generalized to other kinds of tiles and grids. For example, an equally interesting problem replaces the dominos with lozenges, or rhombi, that have the shape of two equilateral triangles glued together. The analogue of a chessboard would be a regular hexagon, or—more generally—any polygonal region with sides of integer length, all parallel to the sides of the regular hexagon. The inside of the polygon would

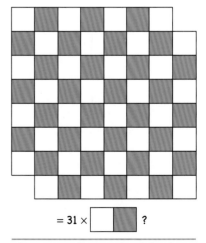

Figure 1. *The puzzle of the dominos and the chessboard with two corners missing.*

be covered by a triangular grid, and the triangles can, if desired, be colored black and white in chessboard fashion. Like a domino on a chessboard, each lozenge covers two units in the grid, one black and one white.

One fact literally jumps out at you when you begin playing with lozenge tilings: a lozenge tiling looks a whole lot like a 2-dimensional picture of a 3-dimensional stack of cubes. (See Figure 2.) In fact, every tiling of a hexagon with sides a, b, c is equivalent to a stack of cubes inside an $a \times b \times c$ box, obeying the rule that no cube can touch air on any one of its three hidden faces.

Percy MacMahon, a British number theorist, in 1915 showed that the number of cube stacks in an $a \times b \times c$ box, denoted by $M(a, b, c)$, is given by an elegant formula:

$$M(a, b, c) = \prod_{i=1}^{a} \prod_{j=1}^{b} \prod_{k=1}^{c} \frac{i + j + k - 1}{i + j + k - 2}.$$

This is a sort of "generalized binomial coefficient." In fact, if $c = 1$, the formula reduces to the standard binomial coefficient:

$$M(a, b, 1) = \left(\begin{array}{c} a + b \\ a \end{array} \right).$$

In the early 1960s, three physicists brought a different perspective to the study of lozenge tilings. One was Pieter Kasteleyn, a physicist who at that time worked for Royal Dutch Shell; the other two, working as a team, were H. Neville V. Temperley, who was then at the Atomic Weapons Research Establishment in England, and Michael Fisher, who was at King's College London.

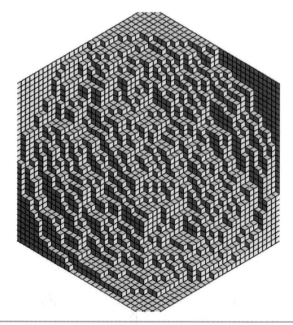

Figure 2. *A random tiling of a hexagon by lozenges. Each lozenge is formed by two equilateral triangles glued together. (Figure courtesy of Richard Kenyon.)*

Kasteleyn, Temperley and Fisher ignored the third dimension and viewed the problem as one of counting *perfect matchings* of a *bipartite graph*. The idea can be seen clearly from Figure 3, which shows a typical domino tiling of a large checkerboard. Each domino in Figure 3a corresponds to two dots in Figure 3b, which lie at the centers of the two squares that make up the domino. The two dots are connected by a line segment that runs down the middle of the domino. The result is a square grid of dots in which every dot is "matched up" with one and only one other dot. This is a perfect matching on a square lattice. Lozenge tilings of hexagons, similarly, correspond to perfect matchings of a hexagonal lattice. If one thinks of the lattice as a crystal in which each dot is an atom, then the perfect matching breaks the crystal down into two-atom molecules or "dimers."

From this point of view, the domino and lozenge tiling problems are special cases of a more general question of how to count the number of perfect matchings of a planar bipartite graph. This is the problem that Kasteleyn, Temperley, and Fisher solved in the early 1960s, by using the *adjacency matrix* of the graph. This enormous matrix has one row and column for each vertex, and contains a 1 in the i-th row and j-th column if vertex i is adjacent to triangle j. Every other entry is 0.

If the white vertices are listed first and the black vertices are listed second, then the adjacency matrix looks like this:

$$K = \begin{bmatrix} 0 & A \\ A^t & 0 \end{bmatrix}.$$

The "0"s represent the fact that vertices of the same color are never adjacent. The matrix is symmetric about its main diagonal because vertex i is adjacent to vertex j if and only if vertex j is adjacent to vertex i.

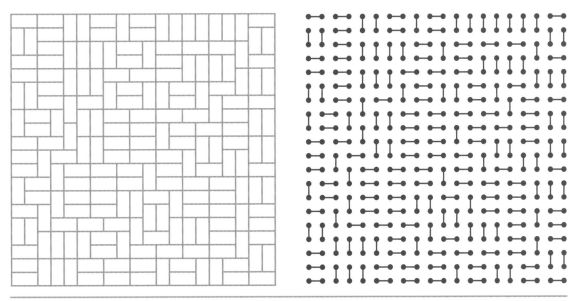

Figure 3. *How to represent a domino tiling as a perfect matching on a checkerboard graph. (left) A domino tiling of a grid. Each domino covers two squares of the grid. (right) The centers of the squares in (the left-hand side) are represented by dots; and two dots are connected or "matched" if their squares belong to the same domino. In a perfect matching, each dot has one and only one match. (Figure courtesy of Richard Kenyon.)*

James Propp. *(Photo courtesy of James Propp.)*

Kasteleyn proved that for lozenge tilings, the number of perfect matchings equals the determinant of A—an even more succinct formula than McMahon's. The reason is that nonzero terms in the determinant of A arise exactly when each white triangle (represented by the i-th row of A) is matched up with exactly one adjacent black triangle (i.e., a column of A such that $a_{ij} = 1$). The determinant simply adds 1 each time a perfect matching is detected.

For tilings other than the lozenge tiling, Kasteleyn's method requires some modification (some 1's in the adjacency matrix have to be changed to -1's) and the results it produces are not quite as beautiful. For example, using this method one can show that the number of domino tilings of the 8×8 chessboard is 12,988,816, or 3604^2. The answer is vaguely unsatisfying because it's not clear where the number 3604 came from. "The formula isn't sweet, the way the generalized binomial coefficient is for counting tilings of the hexagon by lozenges," says Jim Propp of the University of Massachusetts at Lowell.

Propp, along with several co-authors, began working on the domino-tiling problem in the early 1990s. He was convinced that a "sweet" formula for domino tilings must exist, but the chessboard was the wrong region to look at. He noticed an important difference between domino tilings of a chessboard and lozenge tilings of a hexagon: In the latter figure, all of the triangles along any side of the hexagon have the same color. The preponderance of cells of the same color reduces the number of tiling options. A chessboard doesn't have the same imbalance along its edges because white squares alternate with black squares. Thus, Propp came up with the idea of looking at the *Aztec diamond*—his term for a diamond-shaped region composed of square pixels (see Figure 4). Like the hexagon, this region has an imbalance of colors along each edge, although of course it has an overall balance of whites and blacks. (Otherwise, like the mutilated chessboard, it wouldn't be tileable at all.) Noam Elkies and Michael Larsen proved that the number of domino tilings of the $n \times n$ Aztec diamond, represented by $AD(n)$, has as sweet a formula as anyone could hope for:

$$AD(n) = 2^{n(n+1)/2}.$$

It's easy to check, using this formula, that an Aztec diamond has many fewer tilings than a comparably sized chessboard. For example, the Aztec diamond with 60 squares has only 2^{15} or 32,768 tilings, compared to the nearly 13 million domino tilings of the 64-square chessboard.

But it's not just the size of $AD(n)$ that's important—it's the exceptionally simple form. The fact that it's a power of two strongly suggests that a random tiling of an Aztec diamond corresponds in some way to $n(n + 1)/2$ random flips of a coin.

Along with Greg Kuperberg, Propp found a particularly nice illustration of this correspondence between domino tilings and coin flips, which they called "domino shuffling." To construct an Aztec diamond tiling of order n, you start with an Aztec diamond of order 1, which is just a 2×2 square. There are two ways to cover it, either with two vertical dominos or two horizontal ones. Choose one by flipping a coin. Now push the two dominos away from each other. This creates two new 2×2 "holes" that can each be filled in either with two verticals or two horizontals, to create an Aztec diamond of order 2. Again, flip a coin to decide how to fill each one in. Repeat the procedure over and over—expand the diamond, then fill in the holes by flipping a coin. There are some technical difficulties to overcome (beginning with step three, some of the dominos in the expanding colony may annihilate each other), but Propp and Kuperberg showed that Aztec diamonds can be identified in a precise way with sequences of coin flips.

At this stage, domino tilings moved to a new level of sophistication. With the introduction of a random generation procedure, researchers could for the first time move past the simple enumeration of tiling patterns, and ask what a random pattern actually looks like. The answer came as a stunning surprise. (See page 46, "Aztec Diamond.") Instead of a hodgepodge of dominos oriented in various directions, random tilings turned out to have four highly ordered regions in the corners, which Propp called the "frozen regions." Only in the center, within a roughly circular region (the "liquid region"), were the dominos actually free to point in any of the four compass directions. (Here we adopt the convention that a domino "points" from its black square to its white square.)

Now Propp asked a question that no one had thought of before. What would happen if the Aztec diamond was fixed in size, but the dominos were made smaller and smaller? (Technically, this slightly changes the shape of the diamond, as the corrugations along its edges become finer and finer.) In 1996, Elkies, Propp, and Henry Cohn proved the "Arctic Circle Theorem." In a random tiling by infinitesimal dominos, a circle inscribed in the fixed diamond forms the dividing line between two different phases of matter. Outside the circle, the tiling is frozen into a perfectly regular crystalline pattern. Inside the circle, it enters the liquid phase.

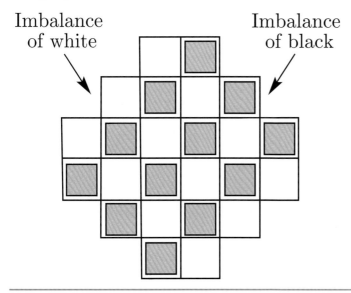

Imbalance of white

Imbalance of black

Figure 4. *In an Aztec diamond, the border has a local imbalance of white or black squares, which reduces the number of tiling options near the border. This helps to explain the "frozen" regions near the border in random tilings of an Aztec diamond.*

This statement is more subtle than it appears because randomness is an essential ingredient. It's *possible* to create a tiling with "unfrozen" tiles in the frozen region, but it's staggeringly unlikely that you will get such a tiling if you choose at random. There is an analogous phenomenon in statistical physics: It's *possible* for all the molecules of gas in a room to spontaneously move to the north side of the room, creating a vacuum on the south side. But it's so unlikely that, in practice, it never happens.

Propp's group also brought a third dimension to domino tilings. (See Figure 5.) In the 1980s, William Thurston showed

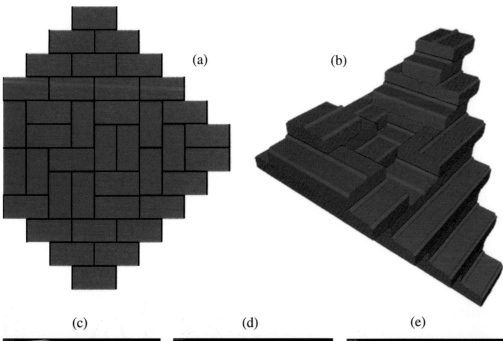

(a)   (b)

(c)   (d)   (e)

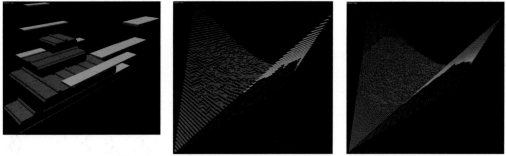

Figure 5. *How to convert domino tilings of Aztec diamonds into random surfaces. (a) A domino tiling of (part of) an Aztec diamond. (b) A 3-d version composed of Lego-like bricks. (Each brick is the size of two dominos). When viewed from directly above, the figure will look exactly like a domino tiling. (c) A random tiling of an order-5 Aztec diamond. The height function forces the Lego-like bricks to lie on top of the gray stepped surface. (d) A random tiling of an order-51 Aztec diamond, converted to three dimensions. The random surface is becoming more apparent, and the gray steps are less apparent. (e) The order-256 tiling from the "Aztec Diamond," p. 46, converted to a random surface. (Figure courtesy of Benjamin Young.)*

that any perfect matching of a two-color tiling pattern creates a well-defined "height function," whose graph is a stepped surface, like the stack of cubes in the lozenge problem. "He showed that it's not just an optical trick," Propp says. "It gives you a new way of thinking about where the obstructions to tileability come from geometrically."

In the case of domino tilings, one can create the stepped surface out of bricks that resemble Lego building blocks (see Figure 5). When seen from above, as in Figure 5a, the arrangement of bricks looks like a tiling by red and blue dominos, with the red dominos pointing one direction and the blue dominos pointing the other. From the side, as seen in Figure 5b, the shape of the stepped surface becomes visible. To visualize domino tilings of the Aztec diamond, you have to imagine placing the Lego-shaped bricks in a V-shaped tray (Figure 5c-e). As the size of the bricks becomes smaller, the stepped surface looks smoother and smoother—and, in fact, the Arctic Circle Theorem implies that there is a well-defined limit surface, which has four planar, sloping facets in the corners, a smooth and gently warped shape in the center, and a circular crease separating the two regions.

Richard Kenyon, of Brown University, Cohn and Propp also showed that a well-defined limit surface exists for *any* polygonal boundary, not just a diamond. In other words, if you fill the boundary up with random domino tilings, and let the dominos get smaller and smaller, the random surfaces defined by these tilings will cluster around a single surface, which with very high probability will have a mixture of flat facets and pieces with smoothly varying slope. The same statement is also true for surfaces defined by lozenge tilings.

Philosophically, this was an astounding fact. To put it succinctly, randomness implies determinism. The shape of the boundary uniquely determines the shape of the random surface with that boundary. However, there was just one problem. Except for the simple cases of the Aztec diamond and the hexagon, Propp had no idea how to figure out what the limiting shape was.

"It was clear to us that there was a much broader realm of questions," Propp says. "We proved a soft analysis result that says there would be a well-defined limiting height function in the general case, but it didn't say how you would find that function constructively. At this point, there was no reason to think it was do-able, and no reason to think it wasn't."

The domino problem was about to take one more quantum leap in sophistication. "I remember as an undergraduate, sitting in a classroom at Harvard when Jim Propp came and started showing us pictures of Aztec diamonds," says Scott Sheffield. "He got all sorts of students looking at Aztec diamonds and domino tilings. At the time a lot of people thought of it as a cute game, a problem you give to undergraduates. It didn't require a lot of knowledge; you could get by with some raw cleverness. But that gradually changed, and it began to be taken much more seriously." So seriously, in fact, that the 2006 Fields Medal—the mathematical equivalent of a Nobel Prize—was awarded to the researcher who first explained what was going on.

Richard Kenyon. *(Photo courtesy of Richard Kenyon.)*

Andrei Okounkov. *(Photo courtesy of Andrei Okounkov.)*

Broadly speaking, the first stage in the evolution of random tilings was existence; the second stage was counting; and the third stage was taxonomy. The fourth and most advanced stage was physics.

By the late 1990s, Propp's group had discovered the third "phase of matter." In certain periodic tilings, a gaseous phase exists along with the frozen and liquid phases. For instance, Figure 6 shows a region that Propp calls a "fortress," which has been tiled with $45° - 45° - 90°$ triangles. When the triangles are paired into dimers, they form two different shapes: larger isosceles right triangles (which are colored light gray in Figure 6), and diamonds (colored dark gray). In the random tiling in Figure 6, you can see three distinct regions. In the striped frozen region, the light and dark gray dimers occur in a rigid pattern. The type of dimer that occurs at any given point is strictly determined by the boundary layer. In the liquid region, nearby dimers are weakly correlated—you can see long strings of dark gray dimers—but the correlation drops off exponentially. The type of dimer that occurs at any given point depends only weakly on boundary effects. Finally, in the center of fortress is a gaseous region, where the dark gray dimers form isolated molecules. The probability of finding a dark gray dimer at any point in the gaseous region is a constant; the "molecules" in this region do not see the boundary at all.

When the tiling is viewed as a random surface, the frozen phase corresponds to a piece of surface with a constant slope—in other words, a planar facet. In the liquid phase, on the other hand, the slope can vary smoothly. That is, the probability of seeing a particular dimer in a particular orientation also varies smoothly from point to point.

From the point of view of random surfaces, the gaseous phase corresponds to planar facets—as does the frozen phase—but the reason is quite different. In the frozen phase, the constancy of the slope of the height function is enforced by a rigid crystalline arrangement of the molecules. In the gaseous phase, it is enforced by probability. The slope of the height function is governed by the proportion of each type of dimer. Because the probability of finding a certain type of dimer at any point within the bubble is constant, the slope of the height function remains constant throughout the bubble.

With three phases that behave so much like the phases of matter, researchers began to deliberately apply the tools of statistical mechanics. In a paper written with Richard Kenyon and Andrei Okounkov, Sheffield showed that each slope of the random surface corresponds to a unique *Gibbs measure*. A Gibbs measure describes the bulk physical parameters of a system comprised of many particles. In a gas, for instance, there may be trillions of molecules, yet the bulk properties boil down to only two: pressure and temperature. A crystal, on the other hand, may have other bulk properties, such as refractive index or net magnetization.

The uniqueness of the Gibbs measure means that in random tilings by infinitesimal dominos, all the properties of the tiling at a point are governed by the slope of the corresponding random surface at that point. It is the ultimate vindication of Thurston's point of view—or even before that, Percy MacMahon's point of view. Not only is the height function not merely

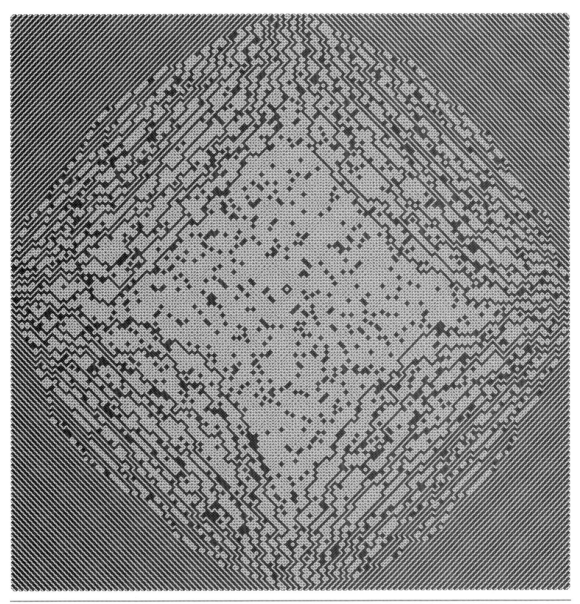

Figure 6. *Diabolo tilings are constructed from two different types of tiles. The dark tiles consist of two right-angled isosceles triangles joined along a hypotenuse, to form a diamond; the light tiles consist of two right-angled isosceles triangles joined along a leg, to form a larger triangle. A random diabolo tiling of the (approximately) square region in this figure displays three different "phases." The corners are frozen; the intermediate region, bounded by a circle, is liquid; and the inner region, bounded by an astroid, is gaseous. (Figure courtesy of James Propp.)*

an "optical trick," but it conveys complete information about the state of the tiling.

But Sheffield's theorem was still only a statement of principle, not a practical tool. It said that the height function contained all the information about the physical state of the tiling, but it didn't explain how to compute the height function.

In a series of papers between 2003 and 2005, Kenyon and Okounkov gave precise and beautiful answers to these questions. First, they converted the probability problem—which limit surface do the stepped random surfaces cluster around?—into an energy problem. For any particular slope, they effectively computed the *surface tension* of the random surface at a point with that slope. Integrating the surface tension gives an energy functional, and the random surface is then the unique surface that minimizes the energy.

For any tiling defined by a perfect matching, the surface tension of the random surface, and therefore the energy functional, is controlled at the microscopic level. Even a 2×2 patch of tiles is sufficient to compute the surface tension function exactly. First Kenyon and Okounkov compute a function of two complex variables, called the *characteristic polynomial* of the tiny patch of tiles—a simple, finite calculation. From this polynomial, they can calculate the precise way that the surface tension depends on the gradient of the random surface. The set of all possible gradients forms a polygon in the tiling plane. Outside this polygon the surface tension is infinite, and so surfaces with gradients lying outside the polygon are impossible. The corners of the polygon of allowable gradients correspond to frozen faces. Gradients that lie inside the polygon correspond (for the most part) to liquid regions.

At this point it may be worth pointing out one difference between mathematics and physics. A physicist might not be too surprised to find out that the surface tension function is determined at the microscopic level. But for a mathematician, possessing a *proof* of this fact makes all the difference. "There are other very simple random surface models for which nothing like this is known," says Sheffield. "There is no nice, explicit formula for the surface tension function."

"It is, basically, a miracle that for certain random surface models, such as dimers, the surface tension is known explicitly," says Okounkov. "For a generic random surface model, so to speak, one may define the surface tension abstractly and prove some basic properties like convexity, but there is little hope to go further."

However, there were even more miracles to come. Once they knew that a random surface minimizes an energy functional, and they could write the functional explicitly, Kenyon and Okounkov could set up a differential equation (called an Euler-Lagrange equation) that describes the random surface. However, it was still far from obvious that the equation could be solved explicitly.

A good comparison would be the Euler-Lagrange equation for the shape of an ordinary soap film. The surface tension function for a soap film is very simple and smooth, and the energy functional (the integral of the surface tension) is even simpler: it's just the area of the soap film. By contrast, the surface tension for a random surface is an oddball function, with sharp corners and a sudden jump to infinity outside the polygon of allowable slopes.

An explicit formula for soap films, called the Weierstrass parametrization, has been known since the 19th century. However, to specialists that parametrization seemed like a special case, dependent on the unusually simple form of the energy functional. It is the last thing one would expect to obtain for random surfaces, with their strange surface tension functions.

"After we wrote down the original differential equation, a nonlinear partial differential equation, we stopped working on the problem because there was no reason to expect to be able to solve it," says Kenyon. But eventually they realized that it was a "disguised version" of another kind of differential equation that was known to be solvable, called Burger's equation.

"What guided us toward this change in variable was some physical intuition," Kenyon elaborates. "If you think not about the typical shape of the random surface but the fluctuations about that shape, you find that in certain problems, the fluctuations have conformal invariance properties. This is something that conformal field theorists like to espouse." (See, for example, "Brownian Motion, Phase Transitions, and Conformal Maps," in *What's Happening in the Mathematical Sciences*, Vol. 6).

"A soap bubble is isotropic—its surface tension is the same in any direction—but dimers have preferred directions in space. The orientation with respect to the underlying lattice is important," Kenyon says. "So we had to make a change of variable to make it look conformally invariant. When we did that, the differential equation turned into Burger's equation. In hindsight, the new variable is the natural variable for the problem."

Okounkov adds, "Our work on Burger's equation began as a stroke of luck . . . . But, in terms of math, it is just an elementary observation about the Euler-Lagrange equation." Their explicit solution to this equation was directly comparable to the Weierstrass parametrization of soap films.

This discovery blew the lid off the random-tiling problem for lozenges. For any given boundary, you could compute exactly where the frozen and liquid regions would fall. And it made a new connection with previously unsuspected parts of mathematics. If the region to be tiled is polygonal, the boundary between the frozen region and the liquid region is an algebraic curve—in other words, a curve described by a polynomial in two variables (see Figure 7, where the boundary curve is a cardioid). This is an amazingly far-reaching generalization of the Arctic Circle Theorem, in which the boundary between the two phases was a circle, described by the polynomial equation $x^2 + y^2 = 1$. It seemed as if the circle arose because of the symmetry of the problem—but in reality, according to Okounkov and Kenyon's theorem, that curve emerged because of its *algebraic*, not geometric properties. Not only did Okounkov and Kenyon find the locus of the phase transition exactly, but they came up with a remarkably simple algebraic technique for finding the slope of the random surface at any point in the liquid region.

This avalanche of beautiful and unexpected results, in addition to his other contributions, earned Okounkov a

This discovery blew the lid off the random-tiling problem for lozenges.

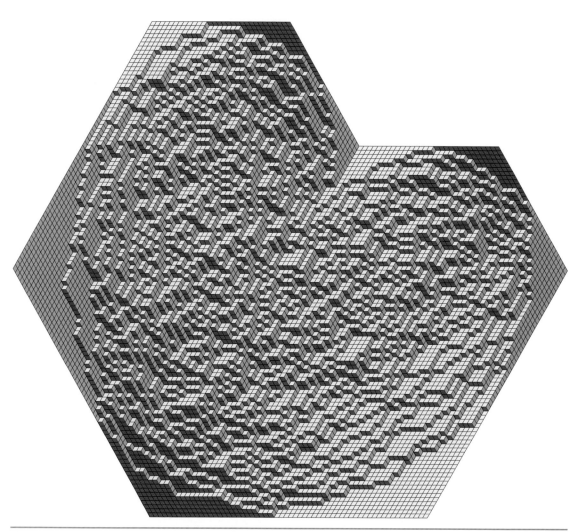

Figure 7. *Okounkov and Kenyon's theorem on random tilings (or perfect matchings), illustrated here for a lozenge tiling. For any random tiling of a region with a polygonal boundary, the solid and liquid phases will be separated by a curve defined by polynomial equations that can be explicitly calculated. Here, the boundary curve is a cardoid. (Figure courtesy of Richard Kenyon.)*

Fields Medal at the 2006 International Congress of Mathematicians in Spain.

What lies in the future for random surfaces and random tilings? Of course, mathematicians will continue to investigate phase transitions in other random surface models, which have not been solved yet. Most of the results described in this chapter apply only to tilings that correspond to perfect matchings of a bipartite graph. However, the phenomenon of phase transitions is surely more general than this.

Physicists will undoubtedly want to explore more realistic models of crystal growth and facet formation, including models in three dimensions. The lozenge and domino models, even though they "look" three dimensional, are best described as "2 + 1"-dimensional, Sheffield says—with two dimensions of the tiling plane coupled to one dimension of the height function. Real engineering problems are "3 + 3"-dimensional, with the three dimensions of space coupled to three dimensions of a stress function that attempts to deform the crystal. In other random surface models, Sheffield says, the surface tension function may not be convex. The beautiful connection between randomness and energy minimization, in which random surfaces congregate around a unique limit surface that minimizes an energy functional, may be broken. There may not be a unique limit surface any more.

Okounkov and Kenyon are continuing to investigate the connections between the dimer model and conformally invariant physical processes, such as Schramm-Loewner evolution (described in *What's Happening in the Mathematical Sciences*, Volume 6). For instance, Kenyon says, any two lozenge tilings can be used to define a set of disjoint chains: Take one lozenge from the first tiling, link it to the two overlapping lozenges from the second tiling, link these two to the next overlapping lozenges from the first tiling, and so on. In the limit as the lozenges get smaller and smaller, Kenyon believes the chains turn into fractal curves with dimension 3/2. This would be true if the chains are conformally invariant, but no one can prove that yet.

In short, random tiling theory has come a long way from the simple brain-teaser of tiling a chessboard with corners removed. "What made the problem change from something recreational to something serious?" Propp asks. "The answer is that there were so many applicable techniques from serious mathematics, from interacting particle systems to calculus of variations. It was a problem that yielded to a bunch of different approaches, but didn't have any trivial solutions."

In short, random tiling theory has come a long way from the simple brain-teaser of tiling a chessboard with corners removed.

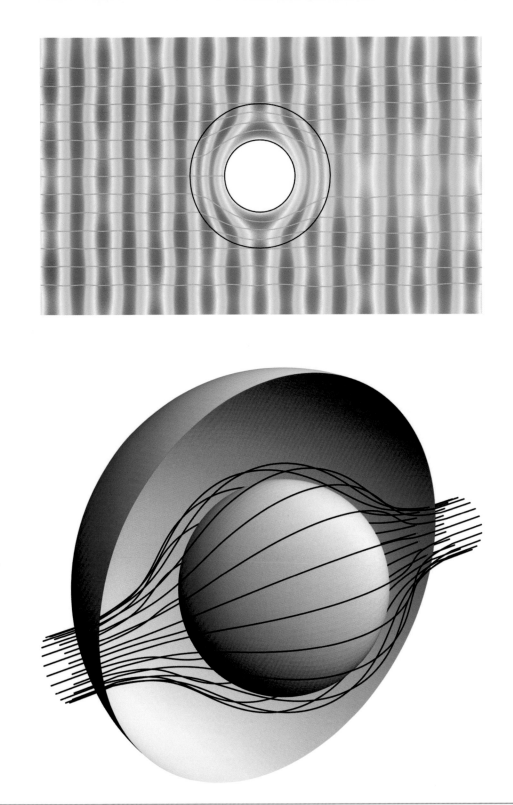

Paths of Light. *The paths of light waves (top) and rays (bottom) through an electromagnetic cloak. (The top image is courtesy of Steven Cummer, Duke University, and the bottom image is courtesy of David Schurig, who is now at North Carolina State University.)*

Not Seeing Is Believing

IN J. K. ROWLING'S MEGA-BEST-SELLING *Harry Potter* series, one of the protagonist's greatest assets is an Invisibility Cloak that he inherits in the first book. It accompanies him through every later book in the series, allowing him to get out of tight scrapes and spy on his enemies. And in the end—spoiler alert!—it turns out to be one of three mythical objects that are the secret to vanquishing death itself.

The part about vanquishing death is pure fiction, of course. But the other properties of the Invisibility Cloak may be closer than you might think to becoming a reality. In recent years, researchers have proved in various ways that invisibility cloaks are theoretically possible, and in 2006 a group of engineers and physicists at Duke University actually built the first rudimentary version of such a cloak (See Figure 1).

Figure 1. *The first working model of an electromagnetic cloak. It is made out of C-shaped copper rings, roughly 3.5 centimeters in width. A conductor placed in the center of the cloak will be more or less invisible to microwaves of one particular wavelength, which is comparable to the size of the rings. (Photo courtesy of David Schurig.)*

> **A disguise makes something look like something else. A true invisibility cloak makes *something* look like *nothing*.**

The technology isn't perfect, of course. The laboratory prototype only renders an object invisible to microwaves, and for only one particular wavelength at that. It is also only a rough approximation to a true invisibility cloak, and mathematicians have proved that even a perfect invisibility cloak would be useless to Harry Potter because it would also be a perfect reflector on the inside. The person inside the cloak would never be able to see out!

"The Harry Potter scenario is not in the cards," says Allan Greenleaf, a mathematician at the University of Rochester. "Cloaking is more likely to have prosaic engineering applications, where you control the frequency of the waves and you want something to be concealed in some way." For example, one of the annoying problems of magnetic resonance imaging (MRI) scanners is that metallic objects have to be removed from the room before they are turned on. If the objects could be cloaked instead, so that the magnetic field could not "see" them, this precaution would not be necessary.

In addition, several researchers have made progress lately on a related problem: hiding an object from sound waves. At present, the materials science technology needed for acoustic cloaking is even less developed than that for electromagnetic cloaking—but the mathematics says that there is no theoretical reason it can't be done.

It is worthwhile, at the outset, to distinguish cloaking from other means of evading detection. For example, the U.S. military has spent billions of dollars on stealth bombers, which are intended to be invisible to radar. However, avoiding radar detection is comparatively easy. A radar antenna "sees" by sending out an electromagnetic wave and recording the reflection. To make a plane invisible to radar, you only have to prevent it from reflecting waves back to their source. A true invisibility cloak, on the other hand, would not reflect waves back in *any* direction—not to their source, and not to anywhere else either. Also, it would not cast a shadow.

Another common way to hide things is to make them take on the characteristics of the background, like a chameleon. But this is a disguise, not invisibility. A disguise makes something look like something else. A true invisibility cloak makes *something* look like *nothing*.

Finally, a third way to hide an object is simply to put it inside a box that's not transparent. Certainly the hidden object is no longer visible. However, the difference between this method and an invisibility cloak is obvious. The object may be hidden, but it's obvious that you are hiding *something*. But when you put the object inside an invisibility cloak, you hide even the fact that you're hiding something.

The story behind electromagnetic cloaking is a complicated one. Three mathematicians discovered the principle, in the context of electrostatics, but had no idea that the materials existed to make their theoretical discovery a reality. Later, a group of physicists, who had been instrumental in developing new classes of materials, rediscovered the mathematics of cloaking as an extreme example of what they call *transformation optics.*

Greenleaf, Matti Lassas, then of the University of Helsinki, and Gunther Uhlmann, of the University of Washington, began working around 2000 on an inverse problem in partial differential equations. It is a modern-day variation on a more classical problem: Given a charge distribution on the boundary of a region in two-dimensional or three-dimensional space, determine the electric field on the inside. If the electrical conductivity of the material inside the region is known (for example, if the region is filled by air or a vacuum), the classical problem leads to a boundary value problem for an elliptic partial differential equation, called the Dirichlet problem, with a unique solution. This is referred to as the *direct* problem.

In the modern variation, known as the *inverse* problem, the electrical conductivity inside the region is unknown. For example, in a medical imaging technology known as electrical impedance tomography (EIT), a voltage distribution is applied to the outside of the body, and a machine measures the current that emerges at points on the surface of the body in response to the applied voltage. Making use of important theoretical insights by a number of mathematicians, a computer then uses this information to reconstruct the conductivity within the body. A region of enhanced conductivity, for example, may represent increased blood flow due to a tumor (See Figure 2, next page).

Because both the electric field and the conductivity are unknown in this problem, the existence and uniqueness of the solution is not covered by the classical theory of differential equations. The problem of uniqueness was named the Calderón problem, after the Argentinian mathematician Alberto Calderón, long associated with the University of Chicago, who called attention to it in 1980.

For a while, research on the Calderón problem followed a path that is perhaps typical for pure mathematics based on real-world problems. Together with John Sylvester, Uhlmann proved the first global uniqueness theorem in 1985, assuming that the conductivity inside the body varied smoothly from point to point. Their result was valid in three (or higher) dimensions, with the two-dimensional theorem being proved by Adrian Nachman in 1996. Various researchers refined the proofs to assume less and less smoothness. But there was a fundamental flaw in all of these theorems; though they were mathematically correct, they did not express very well the conditions that real-world EIT had to operate with. The electrical conductivity of body parts does not vary smoothly at all. One organ may have a different conductivity from another, and the transition between the two may be very abrupt.

It wasn't until 2003 that Lassas, Uhlmann and Greenleaf discovered that singularities are more than just an inconvenience. If the conductivities are allowed to have abrupt discontinuities, go to zero or to become unbounded, the Calderón problem does not necessarily have a unique solution. In other words, two bodies can have different conductivity profiles, yet react to any external electric field in precisely the same way. In fact, it is possible mathematically to design an object that looks to the electric field like...nothing at all! As far as the EIT machine can tell, there is nothing inside it. The object is invisible, or rather, cloaked.

At first, Lassas, Uhlmann and Greenleaf did not realize the significance of their discovery. In medical imaging and other

Gunther Uhlmann. *(Photo courtesy of Mary Levin, University of Washington.)*

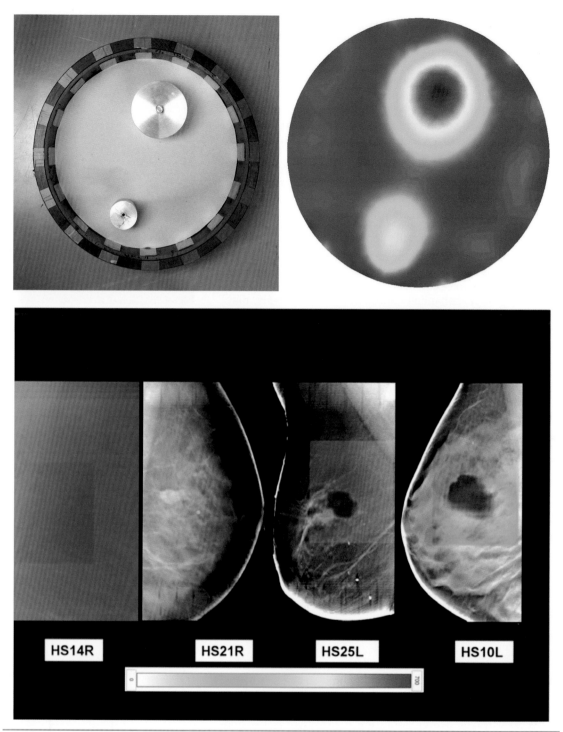

Figure 2. *A laboratory demonstration of electrical impedance tomography (EIT). (Top left) The test subject contains two conducting disks. (Top right) EIT successfully detects the presence of the disks, solely from electrical measurements made around the boundary. (Bottom) In practice, an impedance spectrum is computed at each point, rather than a simple conductivity. Here, processed EIT images (color) are superimposed over mammograms (black and white), as an aid to interpreting the mammograms. Regions with cancer-like spectra are highlighted in red. The two cases at right proved to be ductal carcinomas. (Top images courtesy of the Department of Physics, University of Kuopio, and the bottom image is courtesy of David Isaacson.)*

applications of tomography, the emphasis has always been on *finding* naturally occuring things, like tumors, or cracks in manufactured parts—not on intentionally hiding them. The conductivity profile to make something undetectable is so strange that it seemed very unlikely to be encountered in nature.

There's a saying in computer science: "It's not a bug, it's a feature." Lassas, Uhlmann and Greenleaf were still thinking of invisibility as a bug, a problem to be overcome. It was another team, physicists John Pendry of Imperial College in London and David Schurig and David Smith of Duke University, who first realized that invisibility was a feature...with a future.

According to Pendry, the train of thought that led him to discover invisibility got started in 1996, when he was working with a postdoctoral student, Andrew Ward, on optical fibers. The computer code that calculated the scattering of light by the materials in an optical fiber worked in Cartesian coordinates. However, for a cylindrical fiber it seemed much more natural to transform the equations that describe light—Maxwell's equations—into cylindrical coordinates.

D. Smith, J. Pendry, D. Schurig, and S. Cummer. *(Photo courtesy of John Pendry.)*

Pendry and Ward rediscovered a fact that had been known, in principle, since the nineteenth century: Maxwell's equations are invariant under coordinate transformations. That is one reason why it made sense for Albert Einstein to treat the speed of light as a universal invariant, and to define the shortest path between two points (in a vacuum) as the path that a light ray takes. In Einstein's view, a fundamental constant like the speed of light, or a fundamental geometric quantity like the path of a light ray, should not depend on the coordinates you use to measure it. The whole idea of relativity is that there are no distinguished frames of reference, or coordinate systems, in the universe.

In real materials, rather than a vacuum, Maxwell's laws still remain invariant under coordinate transformations. However, the electrical permeability and magnetic permittivity of the material, which are the coefficients of Maxwell's equations, may change.[1] This again isn't so surprising; many other physical quantities, such as the apparent length of a yardstick or the pace of a clock, change if you move at relativistic velocities.

---

[1] For readers who have forgotten their college physics, the permeability expresses how large an electric dipole moment is created in a material by any given electric field. Similarly, the permittivity expresses how large a magnetic dipole moment is created by any given magnetic field.

But the permeability and permittivity don't just shrink or grow as scalars; they become *anisotropic*. This means that the material may have a different permeability in one direction than in another. Mathematically, the anisotropic permeability and permittivity are represented by functions whose values are matrices (or *tensors*), rather than scalars.

The implication was that physicists could mimic any sort of coordinate transformation, provided that they could design materials with a prescribed, anisotropic permeability and permittivity at any point. In 1996, the importance of this fact was not quite apparent yet, but over the next decade Pendry and David Smith began to discover how to create *metamaterials*. These are precisely engineered structures made from building blocks (you can think of them as "atoms") that, in their early versions, were millimeters to centimeters in size. When you subject them to light waves whose wavelength the metamaterial is designed for, roughly 5 to 10 times the size of the building blocks, the waves will behave in very strange ways. In principle, you can engineer metamaterials to have any desired permittivity and permeability, provided that they are interrogated by electromagnetic waves of the correct wavelength (or frequency).

The first metamaterials that Pendry and Smith developed had a negative index of refraction, a property that is never found in natural materials. Such materials had been conjectured by the Soviet physicist Victor Veselego in the 1960's, but had never been experimentally verified. Negative index materials bend light the "wrong" way (see Figure 3a). The implications of this discovery were already remarkable. For example, Pendry showed that you could make a "perfect lens," which would focus light into regions smaller than a wavelength. You can also make an object located behind the lens appear as if it is in front (see Figure 3b). But Pendry felt that this application only scratched the surface of the possibilities of metamaterials. If you could make light travel along literally any path you choose, surely you could do much more than merely make it bend the wrong way.

"I started asking myself, what can I do with these materials that will really smack people between the eyes?" Pendry says. "My brief was to shock people. That's when I thought, could we make things invisible?"

The idea Pendry came up with was to think of space as a rubber sheet around the object to be concealed. You start with a flat, featureless rubber sheet with a pinprick in it, which represents the way that light waves would move in a vacuum. Then you rip open the pinprick so that the object to be concealed fits completely inside the hole, while the rubber sheet surrounds it. The "ripping open" of the sheet is a coordinate transformation. The waves of light, according to the coordinate transformation, will follow the rubber sheet and will detour around the object (see "The Paths of Light," page 60). Far enough away from the object, the rubber is not distorted at all, and so it will seem to an observer as if the light waves are following exactly the same paths that they would in a vacuum. The object is now cloaked.

According to the transformation laws of Maxwell's equation, the material that enforces this coordinate transformation must be highly anisotropic. Near the inner edge of the "rubber sheet," the permittivity and permeability in the radial direction approach zero, and the refractive index approaches zero. Intuitively, the material forces any incoming electromagnetic waves to bend away from the cloaked object, as if they could not overcome the infinite resistance they are encountering. At the same time, the permittivity and permeability in the tangential directions, along the inner surface of the cloak, remain finite, so that the electromagnetic waves have no problem passing around the object.

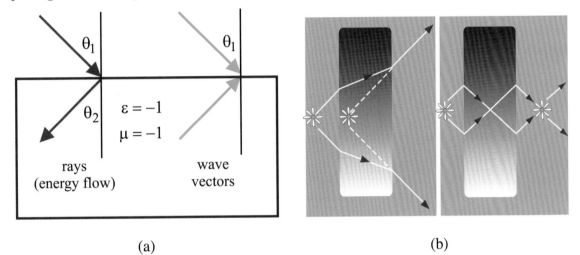

(a) (b)

Figure 3. *(a) When a light ray strikes the surface of a "metamaterial" with a negative index of refraction (blue), it bends the "wrong way." (b) Refraction in an ordinary lens causes an object behind the lens to appear closer. Nevertheless, a viewer still sees it as being behind the lens. By contrast, if the lens has a negative index of refraction, objects behind the lens may appear to be in front of the lens. (Figure 3a courtesy of John Pendry.)*

Pendry first presented his idea at a Defense Advanced Research Projects Agency (DARPA) meeting in 2005, and right away—even at the first meeting—the Harry Potter comparisons started. In June of 2006, he published a theoretical paper in *Science*, with David Smith and David Schurig of Duke, explaining the principle of coordinate transformations and how they could lead to the design of an invisibility cloak. (Ulf Leonhardt of the University of St. Andrews simultaneously published a paper in the same journal with a somewhat different take on cloaking, confined to the ray geometry and only in two dimensions.) By October, Pendry, Schurig, Smith and their collaborators reported in *Science* that they had actually fabricated a metamaterial cloak that rendered a copper disk more or less invisible to 8.5-gigahertz microwave radiation. At that point, the Harry Potter comparisons really began in earnest.

"It was like being hit by a tidal wave," Pendry says. "BBC Today, an early-morning program, called me at about 7 in the morning, and it just went on from there. My last interview was at 10 at night. I didn't know whether I was coming or going! I just wonder what it's like for J.K. Rowling, to have that kind of media attention all the time."

When Greenleaf, Lassas and Uhlmann heard about the papers in *Science*, Greenleaf says, "it psychologically turned our thinking around by 180 degrees." They went back to their

... the mathematicians showed that there are "hidden boundary conditions" at the interface between the cloak and the cloaked region, and these boundary conditions cannot always be satisfied.

2003 work and, in collaboration with Yaroslav Kurylev, now of University College in London, quickly confirmed that they could use the same approach to derive the physicists' results on Maxwell's equations.

On the other hand, there are some important differences between the two groups' results. Lassas, Uhlmann and Greenleaf had no idea, of course, that the materials they were describing were physically feasible. Pendry's group only analyzed what happened to the electromagnetic fields in the metamaterial cloak itself, while Lassas, Uhlmann and Greenleaf considered the complete system of the cloak and the cloaked object. This is mathematically more difficult because the interface between the cloak and the object is a discontinuity. "The coordinate transformation rule for conductivity is basically the chain rule from calculus," Greenleaf says. "But the chain rule only works with smooth changes of variables. Once you have a singularity in the transformation, you have left the world where it applies automatically."

The difference between the two approaches matters when you start trying to cloak an object that is emitting electromagnetic radiation. "It might be somebody with an iPod or a cell phone," Greenleaf says. "If it's radiating at the same frequency [as the waves you are hiding the person from], it might potentially interfere with the cloaking."

In fact, the mathematicians showed that there are "hidden boundary conditions" at the interface between the cloak and the cloaked region, and these boundary conditions cannot always be satisfied. So if you turn on your iPod inside the cloak, or even set off a small electric spark by accident, it is possible that your presence can be detected.

At the same time, Greenleaf, Kurylev, Lassas and Uhlmann have also found a possible solution. At least in the case of a cylindrical object, they showed that you can line the cloak with a grating of a type that had previously been considered by engineers in antenna design. This grating constrains the electric and magnetic fields on the inside surface of the cloak to run in certain directions, and in this way it reduces the number of "hidden boundary conditions" from four to two. That is enough to guarantee that a finite-energy solution to Maxwell's equations on the whole system (object, lining, and cloak) will exist.

However, Greenleaf still doesn't recommend keeping your iPod (or your cell phone) turned on inside the cloak for too long. That's because the cloak will act like a perfectly reflecting surface, and all the energy generated by the iPod will be trapped inside. "Eventually there would be some kind of thermal effect. Presumably, you'd get hot!" Greenleaf says. "This would be bad for three reasons. You'd get uncomfortable, the material might be degraded by heat, and the heat signature might be detectable."

Electromagnetic cloaks also seem to be limited to one frequency, or at best a narrow band of frequencies. When you look at a diagram of the electromagnetic waves passing through the cloak, as in "Paths of Light" (p. 60), it looks as if the waves travel faster in going around the cloak, from one side of the cloaked object to the other, than the more distant, unperturbed waves do. That would apparently violate relativity. But as Veselago already pointed out in 1968, it is only the "phase velocity" of the wave that exceeds the speed of light. There is no violation of relativity because no individual particles are going that fast. However, Veselago showed that when waves of different wavelengths travel through any material, the "group velocity" cannot exceed the speed of light. That means the waves must disperse in the cloak, or travel along different paths, just as they do when light passes through a prism. Thus, if you were wearing an optical cloak, you might be invisible to red light, but you would be perfectly visible in blue and yellow.

At this point, physicists are making progress toward cloaks that work in the visible spectrum. The microwave cloak developed at Duke consists of ten concentric cylindrical Fiberglass sheets, each of which is imprinted with three rows of U-shaped electrical circuits. Each circuit is 10/3 millimeters high and 10/3 millimeters wide. The separation between the cylindrical sheets is $10/\pi$ millimeters, so the "atoms" are roughly cubical. In order to make an optical cloaking device, you would need printed circuits in the 100 nanometer range—which is pushing the limits of current technology, but not out of the realm of possibility.

One of the collaborators on the development of the microwave cloak was an electrical engineer, Steve Cummer of Duke University, who has now gotten actively involved in the search for acoustic cloaking designs. Acoustic waves, of course, do not obey Maxwell's equations, so the first question to ask was whether the equations they obey are likewise invariant under coordinate transformations. "We didn't think so at first," says Cummer.

In two dimensions, actually, there is a perfect analogy between acoustic waves and electromagnetic waves. The electric field is replaced by the pressure (in the medium that the wave is passing through), and the magnetic field replaced by the velocity (of the particles in that medium). With these subsitutions, Maxwell's equations become the acoustic equations, and so cloaking in two dimensions is possible.

However, in three dimensions—the case of greatest practical importance—the analogy breaks down. Graeme Milton, a mathematician at the University of Utah, proved that the full equations of elastodynamics are not invariant under coordinate transformations. At first, this seemed like a death knell for the idea of acoustic cloaking.

Why are elastic waves and electromagnetic waves different? In three dimensions, there are two different kinds of elastic waves. Ordinary sound waves are compressional or *longitudinal*, which means that particles in the wave medium vibrate in the same direction as the wave is traveling. However, elastic media can also support transverse or *shear* waves, in which particles vibrate perpendicularly to the direction of the wave.

By contrast, Maxwell's equations allow only transverse waves, in which the electric and magnetic fields oscillate perpendicularly to the direction of motion. This difference is crucial: the absence of longitudinal waves is what makes cloaking possible. "You can make it work for one or the other, but you can't make it work for shear waves and compression waves at the same time," Cummer says.

However, fluids (such as air or water) do not support shear waves; they only allow compression waves to pass. Thus, within

The absence of
longitudinal waves is
what makes cloaking
possible.

a fluid, Cummer has shown that cloaking should be possible, because there is only one kind of wave to deal with. Fortunately, fluids are the natural environment for many acoustic problems, whether they deal with sound waves in the air or under the water.

Just like an electromagnetic cloak, an acoustic cloak must have some pretty outlandish physical properties. The analogues of permittivity and permeability are the bulk modulus (springiness) and the mass density of the material. To make the cloak work, the mass density has to be anisotropic. But what does this mean? How can a material be "more dense" in one direction than another? That's a real mind-bender.

And yet metamaterials may provide an answer. Milton has suggested the following idea. Imagine a material that is pervaded by bubbles, such that each bubble contains a mass attached to springs in three different direction (See Figure 4). The springs would be constructed with different spring constants, so they would oscillate at different frequencies. The key point is that near a resonant frequency of any particular spring, the bubble would vibrate much more readily, and hence its inertia—that is, its apparent mass—would be decreased in the direction of that spring. But only for motion in that direction; in the other directions, the mass in the bubble would appear as heavy as ever.

Mathematically, this spring-and-mass setup is similar to the miniature U-shaped circuits in the electromagnetic cloak. Their purpose was to change the apparent resistivity or "inertia" of the metamaterial, by exploiting resonance within the circuits.

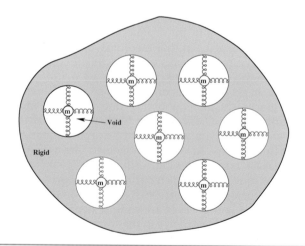

Figure 4. *A schematic view of a metamaterial with anisotropic mass density. Such a material could, in theory, be used to create an acoustic cloak.*

At this point, it is much more difficult to build a bunch of tiny bubbles with springs inside than it is to print circuits onto a piece of Plexiglas. Thus, no working prototype of an acoustic metamaterial has yet been developed. Nevertheless, there is no fundamental reason why it can't be done. The critical questions will be how expensive it is, and how practical. "You don't want to put a 40-meter thick shell around a 10-meter subma-rine," Cummer says. However, he can envision using acoustic cloaks, for example, in architecture. "In the acoustic design of buildings, you have to think about where you put the structural elements," he explains. "Wouldn't it be good if you could just acoustically cloak the beam instead? Then you could decouple aesthetics from acoustics."

Acoustic cloaking has one built-in advantage, compared to electromagnetic cloaking: It doesn't run into any problems with the absolute speed limit, the speed of light. As explained above, that constraint guarantees that any invisibility cloak will be lim-ited to a single frequency, or at best a narrow spectrum. But acoustic waves do not travel anywhere near the speed of light. Thus the problem of dispersion might not be so intractable, and a truly broadband acoustic cloak might be possible. "That doesn't mean we know how to do it!" Cummer cautions.

Cummer's work does not address the problem of cloaking an actively radiating object—in this case, something that is making noise. But Greenleaf and his collaborators' methods apply to this case as well, which is governed by the Helmholtz equation, and they imply that you could cloak a noisy object (for example, your neighbor's drum set?). But as in the electro-magnetic case, the inside surface of the cloak would act like a perfect reflector. "All the acoustic energy would be trapped," Greenleaf says. "On the inside you're going deaf, and on the outside you are undetectable."

Ironically, the first apostle of invisibility now feels as if his idea is distracting people from the full potential of metamate-rials, whether electromagnetic or acoustic. "The cloaking idea was just something I produced to whack people between the eyes," Pendry says. "The trouble is that it seems to have hypno-tized them! What we have here is both a theoretical tool and an experimental means of realizing it, which can send electromag-netic fields any way you please... Using transformation optics, you can imagine gathering all the radiation that hits a sphere into a small volume. This might be useful for communication or energy harvesting."

Greenleaf, Kurylev, Lassas and Uhlmann have recently used transformation optics to describe the permittivity and permeability needed to construct what they call an *electro-magnetic wormhole*, which tricks waves into behaving as though they are travelling not in regular three-space, but on a three-dimensional manifold with a handle attached. As far as Maxwell's equations are concerned, the topology of space has been changed.

Thus it is far from certain that the most important appli-cation of transformation optics will be cloaking devices. Their most important use may not even have been discovered yet. "In a modest way, it's like the invention of the laser," Pendry says. "[The inventors] weren't trying to solve a problem; they just had a great bit of physics. This is a great bit of physics and mathe-matics, too. Of course we know it's likely to have applications. The applications will come later on, by other guys who are the holders of the problem." It may take five years, or ten, or fifty, but eventually metamaterials, and even invisibility cloaks, may become as routine a part of our lives as lasers are today.

Acoustic cloaking has one built-in advantage, compared to electromagnetic cloaking: It doesn't run into any problems with the absolute speed limit, the speed of light.

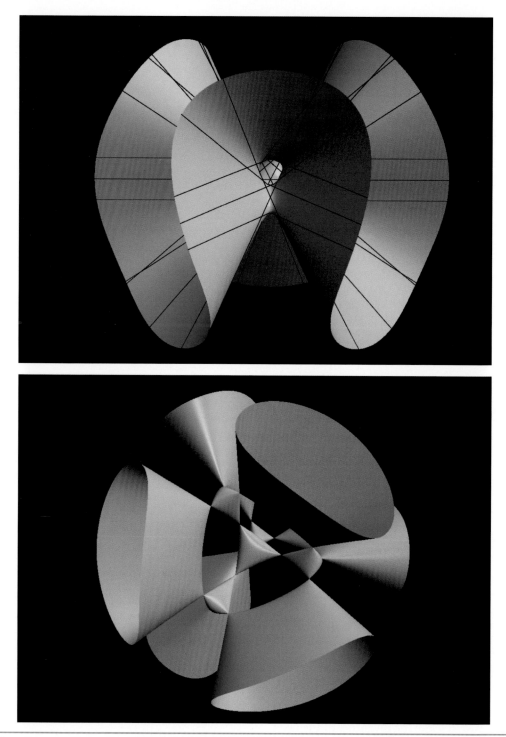

Surface Beauty. *Two examples of algebraic surfaces (two-dimensional algebraic varieties). (top) The Clebsch cubic. (bottom) The Kummer quartic. The Kummer quartic contains multiple singularities, which can be eliminated (or "resolved") by birational transformations. (Figures courtesy of Richard S. Palais for the 3DXM Consortium.)*

Getting with the (Mori) Program

IN 1990, THE INTERNATIONAL CONGRESS of Mathematicians came to Kyoto, Japan, and by a happy coincidence awarded one of that year's Fields Medals—the highest honor in mathematics—to a native son. Shigefumi Mori, a former student at Kyoto University and now a professor at the Research Institute for Mathematical Sciences in Kyoto, earned the award for his completion in 1988 of what became known as "Mori's minimal model program" for three-dimensional algebraic varieties.

The concept of a *minimal model* of an algebraic variety[1] has evolved over the last century, so it is not easy to pin down. But the constant theme has always been a search for the best match between the geometry of a variety and its algebra, expressed by the ring of functions defined on that variety. A minimal model is the most basic, stripped-down version of a variety that still has the same function space.

In the early years of the twentieth century, Italian geometers showed that most two-dimensional varieties, or surfaces, have minimal models. Mori won the Fields Medal for showing that essentially the same thing is true for three-dimensional varieties. However, his work went far beyond a mere extension or elaboration of the earlier results. "Mori's theorems on algebraic threefolds were stunning and beautiful [because of] the totally new features unimaginable by those algebraic geometers who had been working... in the traditional world of algebraic or complex-analytic surfaces," wrote Heisuke Hironaka of Harvard University, in the proceedings of the Kyoto congress. "Three in dimension was indeed a quantum jump from two."

Mori's success led to hopes that the minimal model program could be applied to complex varieties of *any* dimension. Now, in another "quantum jump" of algebraic geometry, Mori's vision has been vindicated. In 2006, Christopher Hacon of the University of Utah and James McKernan of the University of California at Santa Barbara (now at the Massachusetts Institute of Technology) proved minimal models exist in all dimensions, at least for "varieties of general type."

The minimal model program has strong echoes of the geometrization program for three-dimensional manifolds, which was also completed recently (see *What's Happening in the Mathematical Sciences*, Volume 6). According to the geometrization theorem, smooth three-dimensional manifolds can be split into pieces that are either positively curved, flat, or negatively curved. (There are also a few hybrid types that can be ignored for the purpose of the analogy.) The third type represents the

Shigefumi Mori. *(Photographed by Kenji Onishi, The Yomiuri Shimbun.)*

---

[1]As explained below, an algebraic variety is the solution set to a polynomial equation or a system of equations.

> ...three-dimensional *complex* varieties are actually six-dimensional *real* manifolds. And they are not necessarily smooth. (We'll get to that a bit later.) Nevertheless, the similarities are striking.

vast majority of three-dimensional manifolds. Likewise, the minimal model program is a prescription for "cutting away" pieces of a variety that are positively curved, until what's left is either a Fano manifold (the analogue of positively curved), a Calabi-Yau variety (the analog of flat), or of "general type" (the analogue of negatively curved). Again, the third class is the most typical. Strictly speaking, of course, the minimal model program has nothing to do with the geometrization program. For instance, three-dimensional *complex* varieties are actually six-dimensional *real* manifolds. And they are not necessarily smooth. (We'll get to that a bit later.) Nevertheless, the similarities are striking.

Historically, mathematicians have understood algebraic varieties in much the same way that you climb a ladder: one rung at a time. The "rungs" can be listed as follows:

0-dimensional varieties: Carl Friedrich Gauss, 1799

1-dimensional varieties: Bernhard Riemann and Gustav Roch, 1864

2-dimensional varieties: Guido Castelnuovo and Federigo Enriques, 1901

3-dimensional varieties: Shigefumi Mori, 1988

All dimensions: Christopher Hacon and James McKernan, 2006.

Let us climb on the ladder at the bottom.

The Zero-th Rung

A zero-dimensional variety is the set of solutions to a polynomial in one variable:

$$f(x) = a_n x^n + a_{n-1} x^{n-1} + ... + a_1 x + a_0 = 0.$$

(Here we assume that $a_n \neq 0$.) Ever since Gauss proved the fundamental theorem of algebra in his doctoral dissertation in 1799, we have known that such equations always have solutions, provided that the variable x ranges over the complex numbers. In fact, there are always n solutions, but some points may have to be counted more than once (for example, if the polynomial has a factor of the form $(x - \alpha)^2$). Thus any set of n points forms a zero-dimensional variety.

As simple as it is, this example already points out some features that become important in higher-dimensional algebraic geometry. First, the theory always works best when the variables are assumed to be complex numbers. Otherwise, some polynomial equations (such as the equation $x^2 + 1 = 0$) would have "missing" solutions. Also, a point with multiplicity two is algebraically different from a point with multiplicity one (even though they look the same).

The First Rung

The simplest example of a one-dimensional variety, or algebraic curve, is the solution set to an ordinary polynomial in two variables.[2] For example, the equation

$$y^2 = x^3 - x$$

defines a one-dimensional variety known as an elliptic curve. (See Figure 1.)

Notice that the curve in Figure 1 has two ends that "run off to infinity." These two ends can be sewn together by adding in a missing point at infinity. The way to do this formally is to introduce a third variable z and make the polynomial homogeneous (i.e., make all the terms have the same degree):

$$y^2 z - x^3 + xz^2 = 0.$$

Each solution (x, y, z) to the homogeneous equation corresponds to a solution $(x/z, y/z)$ to the original polynomial provided that $z \neq 0$. But in addition, the homogeneous polynomial has a solution $(x, y, z) = (0, 1, 0)$ that corresponds to the "point at infinity" in the original curve. This trick of adding points at infinity turns the original *affine variety* into a *projective variety*, and it is usually assumed either explicitly or tacitly by algebraic geometers.

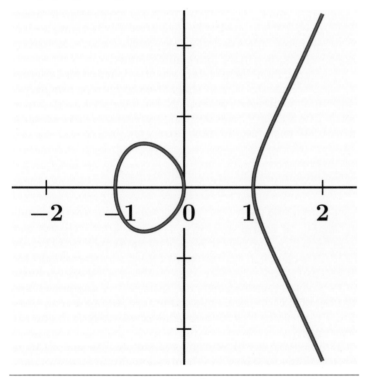

Figure 1. *The elliptic curve $y^2 = x^3 - x$, an example of an algebraic curve (a one-dimensional algebraic variety). (Figure courtesy of Andrew D. Hwang.)*

---

[2]In general, algebraic varieties are defined by systems of polynomial equations. The dimension is, roughly speaking, the number of coordinates required to specify a point in the variety. Thus, in an algebraic curve, one coordinate is enough and two coordinates are too many; that means any two coordinates x_i and x_j are related by some polynomial equation $f(x_i, x_j) = 0$. There may in general be many more than one such defining equation.

Besides omitting the point at infinity, Figure 1 also fails to include any of the solutions in which x and y are complex numbers. Including complex solutions as well as infinite solutions, the solution set to the above polynomial is topologically a torus or "doughnut." Although differential geometers and topologists consider a torus to be a two-dimensional surface, algebraic geometers insist on calling it a curve, because it has one *complex* dimension. For that reason, it is important to know which kind of mathematician you are talking to!

The real virtue of complex algebraic varieties is that they are the most natural setting for extending the theory of functions of a complex variable. Ordinary complex analysis (the kind that students might learn in an undergraduate course) is the theory of meromorphic functions[3] on a complex line. To make the connection more explicit, note that a complex line can be defined inside a complex plane by the equation $y = 0$. (Here the coordinates in the complex plane are assumed to be $x = x_1 + ix_2$ and $y = y_1 + iy_2$, so that two real equations, $y_1 = y_2 = 0$, have been folded into a single complex equation, $y = 0$.) This is the simplest possible algebraic variety, and its meromorphic functions have a particularly simple form: $f(x) = p(x)/q(x)$, where p and q are polynomials. Other varieties have different function spaces. Elliptic curves, such as $y^2 = x^3 - x$, admit a type of meromorphic function called elliptic functions, which were intensively studied in the nineteenth century.

In general, then, the question arises: When do two varieties M_1 and M_2 have the "same" rings of meromorphic functions? This happens precisely when there is a birational transformation between them—a rational function that also has a rational inverse. For example, the affine variety M_1:

$$y^2 = x(x - 1)(x - 2)(x - 3)$$

is birationally equivalent to the affine variety M_2:

$$Y^2 = (1 - X)(1 - 2X)(1 - 3X)$$

because the first curve can be transformed into the second by the change of coordinates $X = 1/x$, $Y = y/x^2$. (The inverse transformation happens to be exactly the same: $x = 1/X$, $y = Y/X^2$.) It is clear that any function of x and y can be written as a function of X and Y, and vice versa. Both maps fail to be defined at the isolated points where $x = 0$ and where $X = 0$, but a finite number of exceptions are allowed for a birational transformation. The varieties M_1 and M_2 are illustrated in Figure 2a and 2b, respectively.

In dimension 1, if the varieties M_1 and M_2 are *smooth*, then any birational map between them is actually an isomorphism. Any apparent exceptional points can be removed. Thus, in dimension 1, each birational equivalence class actually contains one and only one nonsingular projective variety, which is the minimal model. In the example above, M_2 is nonsingular (and therefore is the minimal model), while M_1 is singular. This doesn't seem obvious until you notice that M_1 has four "open

---

[3]In complex analysis, a holomorphic function is defined and differentiable everywhere, and a meromorphic function is a quotient of holomorphic functions. On a complex algebraic variety, a function is meromorphic if it "locally" behaves like a meromorphic function near any point in the variety.

ends" leading to infinity (see Figure 2). After you add in the corresponding point at infinity, the left and right curves will merge into a figure 8, which has a singularity at the crossing point.

In the mid-1800s, Bernhard Riemann and his student Gustav Roch carried the study of the ring of meromorphic functions on 1-dimensional varieties much further. In particular, they studied the vector space of meromorphic functions with prescribed poles at prescribed points (the analogue of the Mittag-Leffler theorem in complex analysis). Their formula for the dimension of these vector spaces established a very close link between the algebra of the function space and topological invariants of the variety.

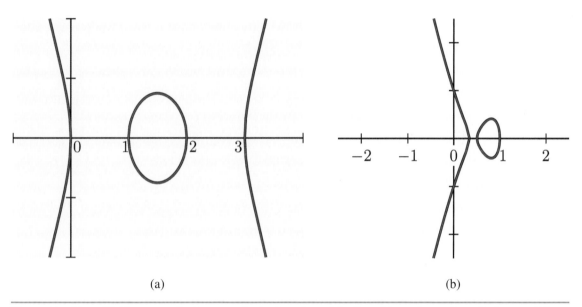

(a)   (b)

Figure 2. *Two elliptic curves that are equivalent under a birational transformation. Only one of them (b) is smooth. The other one (a) actually has a singularity "at infinity." (Figures courtesy of Andrew D. Hwang.)*

A key concept in the Riemann-Roch theorem is the ring of *differentials*, expressions such as dx/y that can be integrated over the variety. For example, the differential dx/y is defined everywhere on the variety $y^2 = x^3 - x$. Its apparent singularity at $y = 0$ is removable because (by differentiating the equation defining the variety) $2y \, dy = (3x^2 - 1) \, dx$, and so $dx/y = 2 \, dy/(3x^2 - 1)$. The right-hand side is still well-defined at the three points where $y = 0$. This differential is an example of what geometers call a *canonical divisor*, a differential that is defined everywhere on a complex variety and (roughly speaking) provides a way to measure volume. The class of canonical divisors (called, not surprisingly, the *canonical class*) would turn out to be a key ingredient in the minimal model program—though its importance would not become apparent until rung three!

From the point of view of the minimal model program, the first rung of the ladder is still a little bit too easy, because any smooth variety is its own minimal model. For more interesting examples we have to ascend to the next rung.

From Veronese's point of view, a 2-dimensional variety in 3-dimensional space was like a shirt crammed into a too-full suitcase. If you give it a little more room, in the form of extra dimensions, you can smooth out the creases and the pinch points.

The Second Rung

Around the turn of the twentieth century, Castelnuovo and his colleague Enriques had the idea of trying to prove a version of the Riemann-Roch theorem that would apply to surfaces, or 2-dimensional varieties. Such varieties arise, for example, as the solution set of a single polynomial equation in three variables. The figures (**Surface Beauty**) on page 72 illustrate two examples of algebraic surfaces in three-space. One of the first things to notice is that such surfaces often contain singularities—points where the surface "pinches off."

In 1881, Giuseppe Veronese—Castelnuovo's teacher— showed that the singularities of a surface in 3-dimensional space could sometimes be resolved by imbedding the surface into more than three dimensions. From Veronese's point of view, a 2-dimensional variety in 3-dimensional space was like a shirt crammed into a too-full suitcase. If you give it a little more room, in the form of extra dimensions, you can smooth out the creases and the pinch points.

The surface imbedded in higher-dimensional space can be considered "essentially the same" as the original one because the two surfaces are birationally equivalent. Castelnuovo and Enriques were the first algebraic geometers to realize the importance of this concept, and to come up with a systematic way to identify the "best" surface in any birational equivalence class—the minimal model. The key concepts involved were the blow-up of a point and the inverse operation, the blow-down of a line.

Figure 3a shows a typical blow-up in real 3-dimensional space, although the reader should bear in mind that this is only a rough representation of what happens in complex projective space. The projection map $p(x, y, z) = (x, y, 0)$ maps a half turn of a helix M_1 to a unit disk M_2. Both of these surfaces are smooth. The function p is rational, and it is one-to-one everywhere except on the z-axis. It compresses the entire z-axis down to a single point. The inverse map does not seem at first glance to be continuous, because of the abrupt jump from the lower boundary of the helix to the upper boundary. However, imagine now that we stretched the helix out "to infinity" on both ends, and then add in the points at infinity, as is customary in algebraic geometry. The effect will be to glue the two ends of the helix together, as shown in Figure 3b, to form a Möbius strip. So in the world of projective geometry, the projection p describes a birational map from a Möbius strip to a disk, in which all the points on the projective line $x = y = 0$ map to a single point $(0, 0)$ in the disk. The line is called the "blow-up" of the point, and the point is called the "blow-down" or *contraction* of the line.

Blow-ups provide one way to generate birational maps between smooth, 2-dimensional algebraic varieties. They also are a systematic way to remove singular points. Castelnuovo proved that they are, in a sense, the *only* way. That is, any birational map can be produced by patching together a sequence of blow-ups and blow-downs. If you start with a variety M that has one of these special Möbius strips in it—which you can think of as a sort of oddly shaped pucker in the fabric of the variety—you can remove the pucker by contracting its central axis to a point and ironing it flat. You may have to repeat this

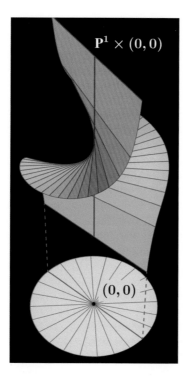

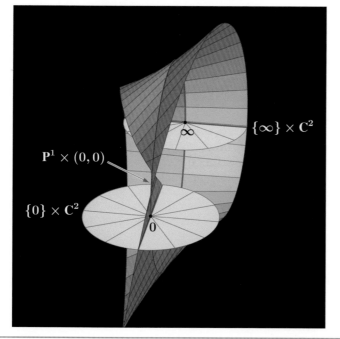

Figure 3. *(top) A blow-up of a point is a birational transformation that maps a disk to a helix-like figure. The inverse map is called a blow-down. (bottom) Adding "points at infinity" turns the helix into a projective variety and glues the two ends together to create a Mobius strip. Blow-ups replace a point in a variety by a (projective) line. Any two-dimensional variety can be reduced to a minimal model by a sequence of blow-ups and blow-downs. (Figures courtesy of Andrew D. Hwang.)*

The *motivation* for finding a minimal model, as noted before, was the quest to prove a version of the Riemann-Roch theorem for surfaces—in other words, to make the geometry match up with the algebra of functions as closely as possible.

procedure several times if your variety contains more than one Möbius strip. But eventually, after all the puckers have been ironed flat, you should have the simplest possible surface that is birationally equivalent to the surface you started with. In other words, you have a *minimal model* of *M*.

Here it is necessary to point out two ways in which the above description is not quite accurate. First, Figure 3 was drawn in the context of real algebraic varieties. For complex varieties, the number of dimensions is doubled. The central axis of the Möbius strip is now a *complex* projective line. (To a topologist, it actually looks like a 2-dimensional sphere. But remember that algebraic geometers consider it to be 1-dimensional.) The analogue of the helix is the *normal bundle* of the projective line. This can be thought of topologically as a fan of planes that are attached to each point of a sphere, such that each plane in the fan points out in two directions perpendicular to the sphere. (Alas, we cannot visualize this object in three dimensions.) The half-twist in the Möbius strip is replaced by an integer, called the *Chern class* of the bundle, which expresses how the normal bundle sits inside the variety. The blow-up of a point has one salient feature: its Chern class is always equal to -1. Conversely, Castelnuovo showed that any "(-1) curve" in an algebraic surface can be blown down to a point.

Thus, Castelnuovo's and Enriques' program for finding minimal models was very simple and explicit. First, find a smooth surface that is birationally equivalent to your variety, by blowing up the singular points. Unlike the 1-dimensional case, that fact alone is not enough to guarantee that your model is minimal. Now you have to hunt for (-1) curves. Blow them down to points. Repeat until there are no (-1) curves left. Then you have a minimal model of your original surface. In most cases, this is the *unique* minimal model in the birational equivalence class.[4]

The *motivation* for finding a minimal model, as noted before, was the quest to prove a version of the Riemann-Roch theorem for surfaces—in other words, to make the geometry match up with the algebra of functions as closely as possible. In fact, Max Noether had made a previous attempt to do this in 1886, an attempt that mathematics historian Jeremy Gray describes as "fundamentally flawed" because Noether did not appreciate the need to work with a minimal model. Only in the context of minimal models were Castelnuovo and Enriques able to prove a useful theorem. Using their generalized Riemann-Roch theorem, they were able to take the first steps toward classifying algebraic surfaces, a project that continues to this day (and will probably never be completely finished).

The Third Rung

While the theory of algebraic surfaces is still an active area of research, there have always been some people who wanted to take the next step up the ladder, to three-dimensional algebraic varieties. The simplest example of a three-dimensional variety is, of course, the solution set to a polynomial in four variables.

---

[4]There is one exception: birationally ruled surfaces, for which the uniqueness does not hold.

As Hironaka noted in his Fields Medal citation for Mori, mathematicians had to make a quantum leap in sophistication to carry out the minimal model program for these varieties. "No one even had the craziness to think you could do it in three dimensions or higher, until Mori," says David Eisenbud of the University of California at Berkeley.

What makes three dimensions so hard? To begin with, it took the experts a while even to reach a consensus on what a minimal model would mean in the three-dimensional case. But the most consistent definition is still the one based on birational maps. A variety M is minimal if any birational map from another variety M' to M either makes M' topologically simpler, or at worst doesn't make it any more complicated.

As in the two-dimensional case, one way to make a variety more complicated is to blow up a point. Thus a variety can be made simpler by the inverse operation, a blow-down or contraction. In a 3-dimensional variety, the blow-up of a point is a complex projective *plane*—that is, a surface, not a curve. Mori showed how to identify surfaces that are candidates for contraction, which are called "extremal rays." These are the analogues of Castelnuovo's (-1) curves in two-dimensional varieties. To return to our earlier analogy, they are surfaces that sit awkwardly inside the variety, creating a pucker in the fabric; they need to be contracted down to a point so that the variety can be ironed flat.

At this point, however, the three-dimensional case diverges from the two-dimensional one. Previously, contracting all the (-1) curves was enough to produce a minimal model. But for three-dimensional varieties, contracting the extremal rays is not enough. The most difficult part, by far—the part that scared off everyone before Mori—is how to deal with *curves* that are embedded awkwardly inside the variety.

From a topological point of view, contracting a surface to a point is called a "codimension-1" surgery. It is comparable to removing a relatively large, benign tumor from the variety. On the other hand, repairing a curve that is imbedded badly is a "codimension-2" surgery. It could be compared to arthroscopic surgery on a human: You make a tiny incision, take out a damaged tendon and sew in a new one that works better. The tools you need for this kind of surgery are more delicate and more advanced.

Mori called his "reconstructive surgery" a *flip*. The name harkens back to the definition of (-1) curves: the surgery removes a bad curve, whose intersection number with the canonical divisor is negative, and exchanges it for a good curve whose intersection number is positive. In other word, it *flips* the sign. Also, in many cases, a flip repairs the "scar tissue" left by the contraction of extremal rays. The contractions sometimes produce singularities that can be smoothed out by a flip. Nevertheless, the minimal variety cannot in general be expected to be smooth (another difference between the two-dimensional and three-dimensional cases). "Algebraic geometers learned to live with these singularities, though their differential geometry is less understood," writes János Kollár of Princeton University.

A final complication was to show that the sequence of surgeries would end, instead of going on forever. In the two-dimensional case, it was clear that any variety has at most

James McKernan.

Christopher Hacon. *(Photo courtesy of Christopher Hacon.)*

finitely many defects that need repairing. Each time you contract a (-1) curve, the topology of the surface gets simpler, until eventually there is nothing more left to do. Similarly, in the three-dimensional case, contracting an extremal ray makes the topology of the variety simpler, and eventually there are no extremal rays left to contract. But flips do *not* make the topology simpler, and it is far from clear that the sequence of flips will ever terminate. To overcome this difficulty, Vyacheslav Shokurov of Johns Hopkins University devised a completely different measure of the "difficulty" of the variety's singularities, and it was this measure that Mori used to complete his program.

As in the two-dimensional case, there is a small and well-understood set of varieties, called Mori fiber spaces, for which the program does not produce a minimal model but does in effect reduce the dimension of the variety by one or more. Although Mori deserves full credit for the key insights in the proof—the existence and termination of flips—the completion of the minimal model program in three dimensions was very much a group effort, including important contributions from Kollár, Shokurov, Miles Reid of Warwick University, Yujio Kawamata of the University of Tokyo, and others.

To the top of the ladder

Even Mori's fabulously ambitious program, completed in 1988, does not work for higher-dimensional varieties. The final assault on the summit of n-dimensional varieties began in the early 1990s, when Yum-Tong Siu of Harvard University introduced new, powerful techniques to study canonical line bundles (a different way of looking at canonical divisors). These techniques had their origin in the field of partial differential equations. In 2003, Shokurov produced a difficult proof of the existence of flips in four dimensions. Hacon and McKernan used Siu's theory to greatly simplify Shokurov's proof, and give a new proof that had no limitation on the dimension n.

There remained the question of the termination of flips. Here Hacon and McKernan, in joint work with Caucher Birkar of the University of Cambridge and Paolo Cascini of the University of California at Santa Barbara, came up with a new approach. "It's still an open question, in dimensions greater than 4, whether the sequence of flips terminates," says Hacon. "Instead of allowing you the greatest freedom in choosing your own sequence, we choose a sequence in such a way that each intermediate step is a minimal model of something, some auxiliary object called a pair, a variety and a divisor." The divisor can be thought of as the "boundary" of the variety. It is added to the variety with a weight parameter between 0 and 1. Hacon and McKernan showed that only finitely many minimal models could exist with parameter values between 0 and 1, and therefore this artfully chosen sequence of flips must terminate. "Everyone was fixated on proving the existence and termination of flips, but we realized we could sidestep this issue by proving just enough of what is needed for an induction," Hacon explains.

As often happens in mathematics, the theorem is not quite as universal as one might like. Currently, it applies only to "varieties of general type," which (as the name suggests) includes most varieties of interest but certainly not all of them. So a challenging problem for future years will be to describe more precisely the boundary between Mori fiber spaces, which do not have minimal models, and varieties of general type, which do have minimal models. Already algebraic geometers have a crude way to tell the difference, called the Kodaira dimension. Mori fiber spaces have the smallest Kodaira dimension $(-\infty)$ and varieties of general type have the largest (the Kodaira dimension is the same as the regular dimension, n, of the variety). Varieties with Kodaira dimension 0, incidentally, are called Calabi-Yau varieties—the analogues of flat manifolds in the geometrization program, as mentioned above. The varieties with Kodaira dimension between 0 and $(n - 1)$ remain, at this point, a mystery.[5] Beyond dimension 4, it has not been proven that they have minimal models, although algebraic geometers suspect that they do. "It looks as if substantial new ideas will be required to complete the picture," says Hacon.

The minimal model theorem may have applications to lower-dimensional varieties as well. For example, the space of all curves of genus g (i.e., curves whose minimal models are topologically a g-holed torus) itself forms a variety of dimension $(3g - 3)$. For genera $g > 22$, this variety is known to be of general type, and thus it has a minimal model. Geometers may learn a great deal by trying to construct this model explicitly. "It will give you a road map for exploring varieties," says Eisenbud.

"It was always one of my dreams to say something meaningful about the geometry of high-dimensional varieties," says Hacon. "As an undergraduate, I had a charismatic professor, Fabrizio Catanese, who introduced me to them." Algebraic geometry has a strong tradition of collaboration between generations—just look at Riemann and Roch, or Veronese and Castelnuovo. Perhaps some student of McKernan or Hacon will be inspired to take the minimal model program to its next step.

> **Algebraic geometry has a strong tradition of collaboration between generations...**

---

[5]Note that there are no Kodaira dimensions between negative infinity and 0. The choice to represent the lowest possible Kodaira dimension as "negative infinity" is to some extent arbitrary; it is similar to defining the dimension of the empty set.

Archimedes Palimpsest. *The inside of the Archimedes palimpsest as it looked in 1998, when the document went up for auction. (Photo Copyright: Owner of the Archimedes Palimpsest.)*

The Book that Time Couldn't Erase

I N OCTOBER OF 1998, A DILAPIDATED OLD BOOK went up for auction at Christie's in New York. Curator William Noel, of the Walters Art Gallery in Baltimore, says it was the size of a box of sugar. Bound in a dark brown leather jacket, the book contained 177 pieces of parchment in the worst conceivable condition: scorched around the edges, worm-eaten in front and back, penetrated through and through by purple blotches of mold. (See opposite page and Figure 1 on page 86.) But in spite of its unpromising appearance, the battered book, now known as the Archimedes palimpsest, has over the last decade opened a new link to the mind of the greatest mathematician of ancient Greece.

The palimpsest[1], or "Archie" as Noel and other researchers call it, had been missing for over 70 years before it resurfaced at the auction in 1998. In spite of its obvious importance, it was unclear who would want to buy a manuscript that was in such deplorable condition. There were only two serious bidders: the Greek ministry of culture and a private American collector, who wished to remain anonymous. "The sale went so fast that you couldn't even keep up with what was going on," says Frederick Rickey, a math historian at the U.S. Military Academy. After a brief flurry of bids, the document was sold—to the American collector, for $2.2 million.

At that point, it would have been natural to expect the palimpsest to disappear from public view, perhaps forever. Instead, over the last ten years it has been subjected to a level of scrutiny whose intensity and openness are unmatched in the history of antiquities. The anonymous owner has spent more than his original purchase price to ferret out the remaining secrets of the document, and to make it available to responsible scholars. Physicists and imaging specialists have literally invented new technologies to study the manuscript— technologies which may now be used to study other ancient manuscripts. Top experts on ancient Greek science have been given the opportunity to interpret the images and debate what they mean. "What was in 1998 an extraordinarily private, illegible product is now free, legible, and public," says Noel.

"Archie" has provided unexpected insights into Archimedes' thinking process. Not only did he anticipate certain ideas of calculus, but he had a surprisingly sophisticated approach to the concept of infinity, and he may have developed some form of the principle of mathematical induction. He also wrote a book on a recreational puzzle in which he may have pioneered some of the ideas of combinatorics. The great scholar apparently had a playful side!

[1]A "palimpsest," as explained below, is a manuscript that has been erased and written over.

> [I]n spite of its unpromising appearance, the battered book, now known as the Archimedes palimpsest, has over the last decade opened a new link to the mind of the greatest mathematician of ancient Greece.

Figure 1. *The Palimpsest as it appeared in 1998. (Photo Copyright: Owner of the Archimedes Palimpsest.)*

Codex C

To read about the history of the Archimedes palimpsest is to find yourself transported into a Dan Brown novel or an *Indiana Jones* movie—with the small difference that it's all true.

The words and diagrams contained in the palimpsest were first set down by Archimedes on papyrus scrolls in the third century B.C. His fame was sufficient to guarantee that his works would be copied and re-copied by scholars for several hundred years. But, as Noel points out, the forces of preservation were in a constant race with the forces of destruction. One after another, the libraries of the ancient world, in Rome, Alexandria, and elsewhere, succumbed to war or neglect. Only one of them—the library in Constantinople—remained sufficiently untouched for Archimedes' writings to survive for the first thousand years. It was here, sometime in the tenth century, that an unknown scribe wrote the copy of Archimedes that has come down to the present. Due to changes in information technology, the scribe was now writing on parchment instead of papyrus; he used iron gall ink, made from the growths ("galls") that form on wasp-infested oak trees; and the parchment was bound into a book rather than rolled up into a scroll.

In 1204, Constantinople's luck finally ran out. The ancient city was sacked by crusaders from Europe, and the library went up in flames. We do not know how many works of Archimedes were lost. We know only that three survived, and these three comprise the sum total of our direct inheritance from Archimedes. Codex A and Codex B made their way to Italy as spoils of war, and Codex C apparently traveled to Jerusalem. Codices A and B were both eventually lost (in 1564 and 1331 respectively), but not before they were translated into Latin. Their permanent place in European culture was thereby assured, as they inspired Renaissance scholars from Leonardo da Vinci to Isaac Newton.

Codex C had a more remarkable fate. It apparently ended up in a monastery in or near Jerusalem, whose monks had no interest in ancient works of mathematics. But the book was valuable for its parchment. "Making a medieval manuscript was tough," says Noel. "You would need to kill about 50 goats to make the amount of parchment in the Archimedes palimpsest." (Parchment was made out of goat skin.) Therefore the monks practiced an ancient form of recycling: They took an old document, scraped away all of the ink that they could, sliced the pages in two, turned them around by 90 degrees and wrote a new document on top of the old one. This was a completely standard procedure back then, called "palimpsesting." (See Figure 2.)

On April 14, 1229—the day before Easter Sunday—a scribe named Ioannes Myronas finished copying a new prayer book. He surely had no idea that the documents he had erased and written over had any value. But the prayer book definitely did have value, for its place and time. It remained in more or less constant use for the next 600 years, preserving its incredible secret. Ironically, instead of destroying Archimedes' work, Myronas probably saved it. History suggests that Codex C would not have survived to the present day otherwise. The proof lies not only in the fate of Codices A and B, but also in the sad story of what happened to Codex C in the twentieth century, after it left its sanctuary in the monastery.

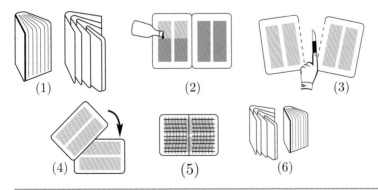

Figure 2. *The palimpsest-making process. The steps are: 1) remove the binding; 2) remove the ink by scraping or treating it with acid; 3) cut the pages in half; 4) rotate them by 90 degrees; 5) write over the erased manuscript (note that the new text is at right angles to the old text); and 6) rebind the new manuscript (which is half as large as the older one).*

Heiberg's discovery
made world-wide
headlines, and his
transcriptions entered
the canon of
Archimedes. But
sometime in the 1920s,
the document itself
disappeared from
Constantinople (now
renamed Istanbul) and
resumed its strange,
wandering existence.

Sometime in the 1800s, the Archimedes palimpsest was transferred back to Constantinople, where it was kept in a library belonging to the Greek Orthodox church. There, towards the end of the century, it began to attract the attention of Western scholars. In 1906, a Danish classicist named Johann Ludwig Heiberg deciphered the faint writing hidden behind the prayers, and identified it as a work of Archimedes. The palimpsest contained parts of several different treatises of Archimedes, some of which were already known from Codices A and B. That is how Heiberg was able to identify the author.

However, the palimpsest also included two works that were previously unknown to modern scholars. In one of them, the *Method of Mechanical Theorems*, Archimedes (in the third century B.C.!) used a method very reminiscent of Cavalieri's Principle from calculus to compute volumes of regions with curved surfaces. This feat must have seemed like magic to ancient mathematicians. Indeed, Archimedes himself considered it his greatest work. According to Cicero, he had a sphere and a cylinder inscribed on his tombstone, preserving for posterity his theorem that the volume of a sphere is two-thirds the volume of the circumscribed cylinder.

In addition to the *Method*, the palimpsest also included the first page of a treatise called the *Stomachion*, about an ancient puzzle that was called the "stomach-ache" because of its difficulty. Due to the brevity and poor condition of the fragment, Heiberg could not figure out what Archimedes was saying, and it remained a mystery to subsequent scholars.

Heiberg's discovery made world-wide headlines, and his transcriptions entered the canon of Archimedes. But sometime in the 1920s, the document itself disappeared from Constantinople (now renamed Istanbul) and resumed its strange, wandering existence. Over the next 70 years, the book suffered more abuse and neglect than it had in the previous 700. (We know this because Heiberg took photos of it in 1906, which can be used for comparison.)

Noel has found evidence that an antiquities dealer in Paris had the book in his possession in the 1930s. We can only surmise what dire circumstances the dealer, who was Jewish, experienced during the Nazi occupation of Paris. He apparently resorted to creating forgeries of medieval paintings on the parchment, perhaps in order to sell them. Three of the pages that Heiberg photographed have disappeared. Four more pages have been painted over with copies of medieval icons of St. John, St. Matthew, St. Luke and St. Mark (see Figure 3). But they were not removed, because the dealer must have found someone willing to buy the whole book.

However, the new owner either could not or did not know how to take care of an ancient manuscript. Parchment has only two enemies, according to Noel: fire and mold. Codex C had survived fire, but now, stashed somewhere in a damp Paris cellar, its pages began to be digested by mold.

In this condition, ravaged many times over—singed by fire, erased and written over by a scribe, fingered by generations of monks, painted over by a desperate forger, and finally grown upon by mold—the Archimedes palimpsest reached Noel's desk in January of 1999, three months after the auction. The

Figure 3. *A forgery painted over the original palimpsest page.*
(Photo Copyright: Owner of the Archimedes Palimpsest.)

new owner entrusted Noel with the job of finding people who
could undo the ravages of time.

Archimedes Meets the X-ray

At this point, for the first time, modern science enters the story
of the palimpsest. Noel, with the owner's approval, chose two
teams of imaging specialists to photograph the fragile document. The teams eventually merged into one: Roger Easton of
the Rochester Institute of Technology; Keith Knox, who was
then at Xerox Corporation in Rochester; and William Christens-
Barry, of Johns Hopkins University.

At first, the imagers tried to create digital images of the palimpsest and process them in such a way that the prayer book text would disappear, leaving only the Archimedes text visible. To do this, they obtained spectra of the document in several different wavelengths of light. The ink used for the original Archimedes document had a slightly different spectrum from the ink used for the prayer book, and both were different from the spectrum of the parchment. With a "linear mixing model," they estimated statistically how much each component—overtext, undertext, and parchment— contributed to each pixel. By subtracting off the overtext and the parchment components, they could reproduce a reasonable facsimile of the undertext all by itself.

It would have been poetic justice if Archimedes had been rescued by mathematics. Alas, reality failed to follow the script. The images that Knox and his colleagues produced were beautiful—but useless for two reasons. First, digital imaging technology was still in its early years. The digital camera, vintage 2001, produced images with a resolution of only 200 pixels per inch. Not only that, the red and green pixels did not match precisely because of chromatic aberration, and this mismatch degraded the fidelity of the processed image. The classical scholars who were trying to read the text complained that it was too blurry.

Secondly, the method didn't just erase the overtext; it erased anything underneath it. Thus, if there was a gap in the undertext, the scholars could not determine whether the gap was real (i.e., nothing was written there) or whether the undertext had simply disappeared behind the overtext. As Easton says, "We did our job a little too well." What the scholars actually needed was a way to see *both* texts, but tell them apart easily.

The final solution that Easton, Knox, and Christens-Barry devised was less mathematical, but it was quick and effective. They photographed the document in two wavelengths— infrared and ultraviolet. The Archimedes ink is completely transparent under infrared light, but the prayer book ink is opaque. Both inks are opaque in the ultraviolet. The imagers assigned the infrared signal to the red channel in a red-green-blue (RGB) plot, and passed the ultraviolet signal to the green and blue channels. In the resulting false-color image, the prayer-book text looks black because it is opaque in all three bands, but the Archimedes text looks red. (See Figure 4.) The classical scholars were now ecstatic because they could see when the Archimedes text was disappearing behind the prayer book, and when there was no text at all.

Knox, Easton, and Christens-Barry have now created false-color images of the entire palimpsest, which became publicly available on the Internet after October 29, 2008—the tenth anniversary of the auction. They have not, by the way, given up on their initial approach of subtracting out the overtext (they call these images "sharpies"). In 2006, they took a new series of images at 11 different wavelengths, using a higher-resolution camera and better light sources (light-emitting diodes, which provide tighter control over the wavelengths emitted). The new sharpies are less blurry and noisy than the old ones, and some scholars even prefer them to the false-color images.

Meanwhile, the problem of the forged paintings remained unsolved. No amount of image processing could reveal what lay underneath the gold paint on those four pages. In 2004, the owner gave his blessing to Uwe Bergmann of Stanford University to attempt a radically new method called X-ray fluorescence spectroscopy.

X-ray fluorescence spectroscopy is really not an imaging technique but an assaying technique: It tells you how many atoms of what kind are found in a certain tiny region of space. For instance, Bergmann is using it to study the configuration of manganese atoms during the chemical reactions of photosynthesis. The idea of using it to image a large object—and a fragile one—was completely untried.

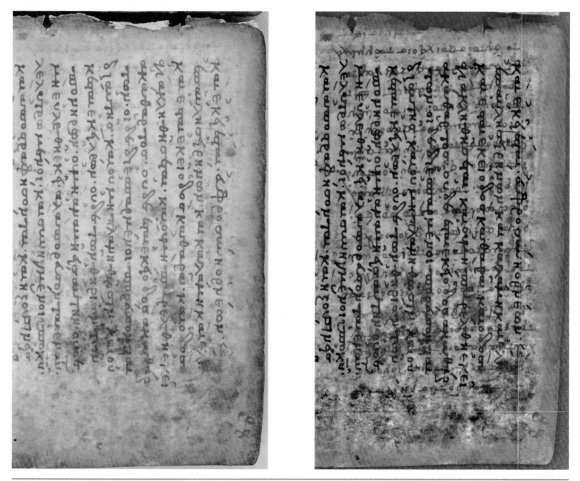

Figure 4. *Ordinary photograph and false color photo of same page. (Photo Copyright: Owner of the Archimedes Palimpsest.)*

Bergmann proposed to map out, pixel by pixel, where the iron atoms on the pages of the Archimedes palimpsest were. Both the original scribe and the prayer book scribe used iron-gall ink, so both inks should be visible, provided that the paint in the forgeries didn't contain any iron. (For the most part, it didn't.) The process would be painstakingly slow because the X-ray beam could only assay one point at a time. Even at a rate of one millisecond per pixel, it would still take 15,000 seconds or more than 4 hours to image a single page. (In reality, it took Bergmann's team about 12 hours per page.)

To acquire an XRF measurement in a millisecond, you need a lot of photons. The number of photons corresponds to the brightness of the X-ray beam, and according to Bergmann, you need a beam one billion times brighter than sunlight. There are only a few machines in the country capable of producing such an intense flux of X-rays, and Stanford's synchrotron is one of them. It is a ring 240 meters in diameter, in which electrons are accelerated to nearly the speed of light. Just as a car's tires squeal when it goes around a tight corner at high speed, the electrons "squeal" as they circle the synchrotron, and the squeals come out in the form of X-rays.[2]

With XRF spectroscopy, Bergmann could watch the text underneath the paintings emerge, one line at a time. "The images were sent around the world in minutes, and within hours we were already getting the scholars' comments," Bergmann says.

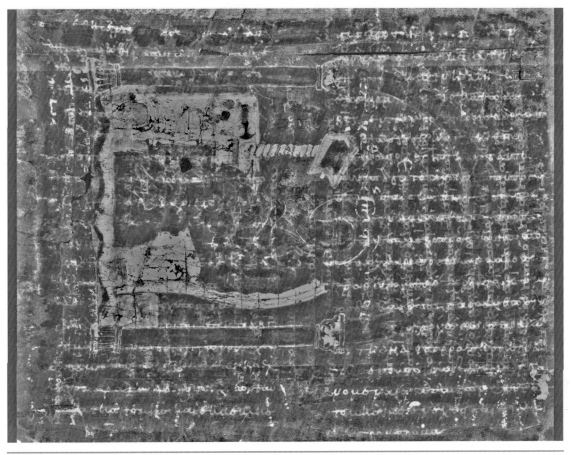

Figure 5. *XRF recovery of the text behind the forgery. (Photo Copyright: Owner of the Archimedes Palimpsest.)*

[2]Here is a less poetic explanation: In order to travel in a circle, the electrons must accelerate toward the center of the circle. According to quantum physics, any accelerating charged particle will emit photons. Because of the extremely high speed of the electrons, the photons are Doppler shifted to a very high frequency—and thus they fall in the X-ray part of the electromagnetic spectrum.

The results were nothing short of miraculous: the letters behind the gold-foil paint were crisp (see Figure 5). Even on the most heavily damaged pages, pages that were nearly black, the X-ray images came out crystal clear. The last page, which was previously illegible, turned out to contain the signature of the scribe who wrote the prayer book—Ioannes Myronas, emerging from anonymity after all those years.

If only Archimedes had still been around! He was, of course, one of the first scientists to experiment with optics. His "burning mirrors" (parabolic mirrors designed to focus sunlight onto enemy ships) supposedly saved the city of Syracuse from Roman attack. Now a different kind of burning ray had proven instrumental in recovering his words. "I think he would be completely thrilled," Bergmann says.

A Plethora of "Plethora"

While the imaging specialists were doing their job, a team of three classical scholars took on the challenge of deciphering the new images. Reviel Netz of Stanford, Nigel Wilson of Oxford University, and Natalie Tchernetska of Cambridge University were not certain at first that any new understanding would come out of the palimpsest. After all, Heiberg had done a seemingly definitive transcription nearly a century ago, when the palimpsest was in much better condition. Were there any more secrets to be discovered?

The answer turned out to be a resounding yes. Heiberg had followed a long scholarly tradition of ignoring the diagrams in the Archimedes codex, assuming that they were corrupted in transmission. In fact, that is true if you look at contemporary editions. Nineteenth-century editors heavily modified Archimedes' diagrams. However, Netz has found strong evidence that the pictures were *not* edited between 200 BC and 900 AD. In particular, Archimedes' diagrams reflect an attitude toward pictures that was quite different from modern geometry books. They were deliberately drawn in such a way that they did *not* necessarily look like the objects they were intended to represent.

Consider, for example, Figure 6 (page 94), from Theorem 14 of the *Method*. (This image comes from the XRF spectroscopy.) Archimedes is trying to compute the volume of the lipstick-shaped solid formed when you cut off a cylinder by an inclined plane. Figure 6 illustrates one step of his argument. To a modern reader, it looks like a triangle nested inside a circle inside a square. But what it really represents is a *parabola* nested inside a circle inside a square. Archimedes deliberately drew the parabola as a triangle! As Netz says, he "threw out the pictoriality" of the picture because he did not want the picture to contain extra information that was not part of the argument. Because Heiberg did not understand Archimedes' picture, he misinterpreted the argument.

More shockers were to follow. The proof of Theorem 14 involves an intricate argument that we would recognize today as a close cousin to integration. Archimedes shows that the areas of certain cross-sections of the lipstick-shaped figure and the corresponding cross-sections of the cylinder form the same ratio as the lengths of certain pairs of line segments. He then "integrates" the cross sections to draw a conclusion about the

[Archimedes'] "burning mirrors" (parabolic mirrors designed to focus sunlight onto enemy ships) supposedly saved the city of Syracuse from Roman attack. Now a different kind of burning ray had proven instrumental in recovering his words.

volumes. However, Heiberg could not read Archimedes's explanation of this step—and that was even before the palimpsest went through a century of abuse.

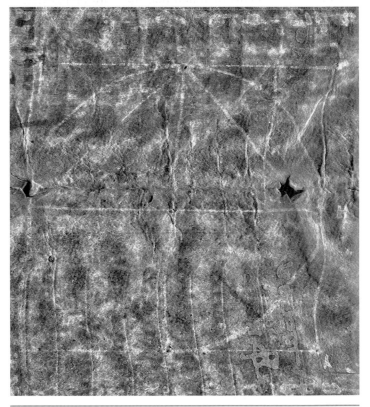

Figure 6. *Archimedes' non-literal diagram. (Photo Copyright: Owner of the Archimedes Palimpsest.)*

With the false-color images, Netz was able to discern the word *plethos*, or "multitude." In Netz's reading, Archimedes is saying that each one of the multitude of line segments corresponds to one of the multitude of cross-sections. The language is amazingly similar to modern set theory; in essence, Archimedes is setting up a one-to-one correspondence, which is exactly the modern notion of when two sets have the same "multitude."

The same word *plethos* turned up again in the one-page fragment of the *Stomachion* (see Figure 8, page 97), which Heiberg hadn't been able to decipher at all. Again, this one word, *plethos*, gave Netz a clue what the document is all about. Netz already knew that the "stomach-ache" puzzle consists of 14 pieces that fit together to form a square (see Figure 7). For most people, the object of the game was to create interesting representational figures out of the puzzle pieces, such as elephants or warriors. But not for Archimedes.

If Archimedes was writing about a *plethos*, he must have had in mind a multitude of something...but of what? Netz believes that he was writing about the number of different ways to fit the pieces *back into the original square*. Mathematically, this is a much more interesting question than how to make an image of an elephant or a warrior. Netz put out a call to mathematicians to see if they could find the answer, and two of them answered.

Persi Diaconis (see "The Fifty-one Percent Solution," page 35), working with paper and pencil, and Bill Cutler, working with a computer, independently found that there are 17,152 solutions. They fall into 536 families of 32 solutions apiece that are related by elementary geometric operations (rotation and reflection).

There is no way to know whether Archimedes computed the answer correctly, or what method he used—the text is too fragmentary, and we only have the first two paragraphs to go on. But even understanding what problem he was writing about is a great step forward. It is the first example known in Greek mathematics of a problem in combinatorics, the theory of counting. The problem has the right feel to it. It involves geometry, certainly a strong point of Greek math; it is not so easy that Archimedes could have solved it effortlessly; but on the other hand it is not so hard that it would require a modern computer or modern-day mathematical techniques.

Thus, the Archimedes text revealed more about the man and his methods than anyone expected—and there may be more. The X-ray fluorescence images revealed a new line of text, unseen by Heiberg, which suggests Archimedes knew a version of the principle of mathematical induction. This principle is yet another tool that modern mathematicians have developed to deal with the tricky concept of infinite sets. Previously, historians believed that Greek mathematicians shunned the concept of infinity. Now it appears that Archimedes, at least, had some very sophisticated ideas on how to deal with it. "Archimedes has tamed infinity to his, if not to our satisfaction. That is more than a footnote, that is a real change to the history of math," Netz says.

Figure 7. *The Stomachion puzzle.*

Archimedes in the 21ˢᵗ Century

So far, the great majority of research on the Archimedes palimpsest has been sponsored by the anonymous owner. As the palimpsest passes the tenth anniversary of its auction, that will change. All of the images will go up on a public website, http://www.archimedespalimpsest.org, from which they can be downloaded by any mathematician or historian who wants to take a crack at deciphering them. Eventually, Netz hopes to complete his own translation of Archimedes, so that the palimpsest can be read by people who do not know ancient Greek.

Meanwhile, the project has opened many new lines of research. Roughly one-third of the prayer book was written over documents that were *not* composed by Archimedes. One of them has now been identified as a collection of speeches by Hyperides, considered one of the great orators of ancient Greece (a rival of Demosthenes). Some of these speeches were not previously known. Another previously unidentified fragment may be Alexander of Aphrodisias' lost commentary on Aristotle. Identifying the authors and subject matter of the remaining parts of the palimpsest will continue to pose a challenge for classical scholars.

On the imaging side, the possibilities for pseudocolor and X-ray fluorescence imaging of other ancient documents seem unlimited. Many ancient documents have been preserved through palimpsests, and could perhaps be reconstructed more accurately with modern techniques. "We want to build a new institute at Stanford called Littera, to read hundreds of cultural objects with X-rays," says Bergmann.

It's a reasonable question to ask whether the documents that were hidden away in the monastery, such as Archimedes' *Method*, really affected mathematical history. After all, they were not available to Newton, Leibniz, and their successors, during the crucial period when calculus was being discovered and its principles worked out. But Netz argues that Archimedes' *Method of Mechanical Theorems* did have an effect *in absentia*. Mathematicians knew from the ancient stories that Archimedes could find volumes of curved figures, and they certainly suspected that he had a secret method for doing this. Now we know that Archimedes was not as methodical as they thought—his theorems, though highly ingenious, did not have the generality of Newton or Leibniz's. So perhaps what Newton and Leibniz didn't know actually helped them! "If they had had the *Method*, perhaps they would have been less eager to look for a secret," Netz says.

Figure 8. *A photograph of the only extant page of the Stomachion manuscript. Can you read it? (Don't forget that Archimedes' text is written horizontally!) (Photo Copyright: Owner of the Archimedes Palimpsest.)*

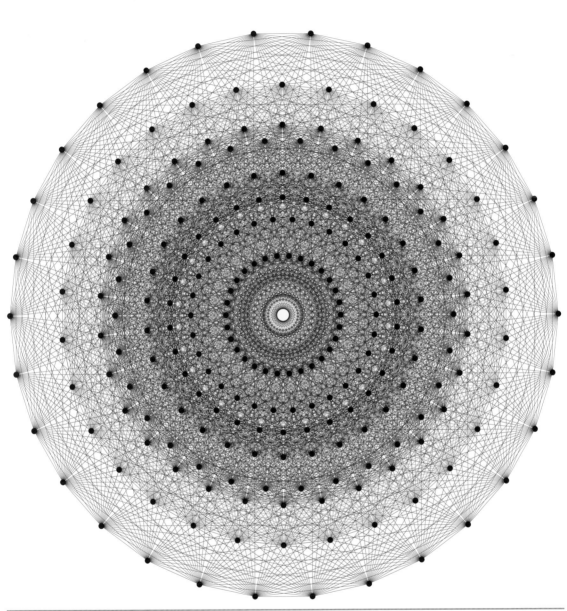

A Symphony in 2, 3, and 5. *The root system for the exceptional Lie group E_8 forms a polytope in 8-dimensional space. The polytope has 240 vertices, and generates a lattice that is the densest possible in 8-dimensional space (in the sense that each lattice point has the largest possible number of "neighbors"). In this picture, the root system has been projected down to a 2-dimensional plane; the plane has been chosen to highlight the 2-fold, 3-fold, and 5-fold symmetries of the root system. Thus the entire picture has 30-fold rotational symmetry. (Figure courtesy of John Stembridge, University of Michigan.)*

Charting a 248-dimensional World

M OST MATHEMATICIANS PROBABLY SEE THEMSELVES—at least in their fantasies—as explorers, questing for new theorems in the same way that the sailors of old quested for new lands or trade routes. But for every explorer, there also needs to be a mapmaker. The cartographers of the sixteenth century, when they drew the first tentative outlines of America on their maps, were permanently engraving the New World onto the psyche of the Old. That knowledge turned out to be more valuable than any of the riches that the explorers hoped to bring back home.

In mathematics, too, there are activities that could be compared to mapmaking, such as compiling data or working out examples that might serve as a guide to other mathematicians. Such contributions usually attract less attention than new theorems, even though they are just as essential for the discovery process.

But in 2007, the usual order of celebrity was reversed. A single calculation, known as the character table for the group E_8, attracted so much publicity that it was chosen as one of the top 100 "science stories of the year" by *Discover* magazine. It was not a new theorem. In fact, the theorem that this calculation was feasible (in principle) had been proven two decades earlier. But its actual completion was comparable to the publication of a new map. It allows the exploration in detail of "representations" of E_8, which are the main ways that the group manifests itself in other branches of mathematics and physics.

The "map" of E_8 was actually only one page out of an atlas, called the Atlas of Lie Groups[1], which is still in the process of being assembled. The Atlas is a combined project of about twenty mathematicians, led by Jeffrey Adams of the University of Maryland. Its stated goal is to make information about the non-compact forms of all Lie groups easily accessible to the mathematical public. Number theorists and physicists (who are some of the most frequent users of Lie groups) especially want to know the *unitary representations* of Lie groups. At present, it is a black art to determine whether a representation is unitary. When the Atlas is finished, that should become a routine computation.

Lie groups in general describe the most symmetric objects in mathematics—objects that have not only discrete symmetry but also continuous or smooth symmetry. Discrete symmetry operations are like reflections in a mirror; they have no intermediate stages. Smooth symmetry operations, on the other hand,

> [I]n 2007, a single calculation, known as the character table for the group E_8, attracted so much publicity that it was chosen as one of the top 100 "science stories of the year" by *Discover* magazine.

---

[1] Lie groups were named after the Norwegian mathematician Sophus Lie, who first studied them in the 1870s. The name is pronounced "lee."

are like the rotations of a sphere; they gradually move the object to its new orientation. Technically, a Lie group is a group (an algebraic structure with one operation) that also has the structure of a smooth manifold. The interaction of three different types of structure—algebraic, geometric, and analytic (i.e., involving the notions of differentiation and integration)—makes the theory of Lie groups extraordinarily rich.

Why should one page of the Atlas of Lie Groups be so much more important, or at least so much more celebrated, than the rest? The answer lies in the special nature of E_8 itself. Like molecules, Lie groups can be divided up into "atoms" of symmetry that cannot be reduced any farther. These atoms, or simple Lie groups, form a sort of periodic table. There are four infinite families, known as A_n, B_n, C_n, and D_n. (The index n, known as the *rank* of the group, can be any positive integer.) Besides these, the table also includes five *exceptional* Lie groups that don't fit into any of the above families: G_2, F_4, E_6, E_7, and E_8. William Killing, a German mathematician, discovered his classification between 1888 and 1890. Like the periodic table when Mendeleev first discovered it, Killing's classification was not complete: It was more of a framework with a few blank spots that had to be filled in later. Killing could not construct the groups E_6, E_7, and E_8, which were first exhibited by Élie Cartan in 1894.

The Atlas of Lie Groups team, *photographed at the American Institute of Mathematics in 2004. From left to right, Peter Trapa, David Vogan, Jeffrey Adams, Fokko du Cloux, John Stembridge, Susana Salamanca, Tatiana Howard, Siddhartha Sahi, Wai Ling Yee, Annegret Paul, Dan Ciubotaru, Dan Barbasch, William Casselman, Alessandra Pantano. (Photo courtesy of David Vogan.)*

As the last of the exceptional groups, E_8 is unique in several ways. First of all, it is the largest. It is 248-dimensional, which means that 248 parameters or variables are required to specify any single element of the group. To compare this with some simpler examples, the smallest Lie group—the set of rotations of the plane, denoted SO(2)—is one-dimensional because any rotation can be described by a single number θ (the angle of rotation). The smallest compact simple Lie group, SO(3) or the set of rotations of three-dimensional space, is three-dimensional. That is because any rotation of space is described by three parameters: the latitude and longitude of the axis of rotation, and the angle of rotation about that axis.

Not only is E_8 larger than the other exceptional groups (which have dimensions 14, 52, 78, and 133 respectively), but it also *contains* all of the others. It is, one might say, the wellspring or mother lode of exceptionality. It also contains quite a few of the non-exceptional Lie groups.

Finally, E_8 is connected to a whole host of intriguing phenomena that occur in eight-dimensional space and nowhere else. For example, seven-dimensional spheres in eight-dimensional space pack together in a uniquely efficient arrangement called the "E_8 lattice." If one sphere in this lattice is centered at the origin, there are 240 spheres touching it—the snuggest possible fit of seven-dimensional spheres. The centers of all 240 spheres form a polyhedron called the *root system* of E_8. The figure, **A Symphony in 2, 3, and 5**, page 98, shows what this eight-dimensional figure looks like when projected down to two-dimensional space. This figure has become virtually emblematic of E_8, but it should be remembered that it conveys only a hint of the group's multidimensional symmetry.

These marvelous properties account for the somewhat reverential way that people who know about E_8 speak about it. John Baez, a mathematician at the University of California at Riverside, calls it "awesome." Physicist Garrett Lisi calls it "the most beautiful structure in mathematics." Mathematician Bert Kostant, of the Massachusetts Institute of Technology, has described it as "a symphony in 2, 3, and 5." The **Symphony in 2, 3, and 5** figure gives a hint of this beauty because it is symmetric under 2-fold, 3-fold, and 5-fold rotations. In eight dimensions these symmetries would be disentangled from one another more clearly.

For many purposes, the most important data about a Lie group are its *representations*, which were briefly alluded to above. A representation of a Lie group is a vector space on which it acts as a group of linear transformations (or matrices). For example, elements of SO(3) are linear transformations of three-space and therefore can be represented as 3×3 matrices of real numbers. Most Lie groups have some representations with lower dimension than the group itself. For example, G_2 has a 7-dimensional representation and thus can be viewed as a group of 7×7 matrices. But the smallest representation of E_8 is 248-dimensional. Therefore, to see even a single element of E_8 you have to write out a 248×248 matrix.

One reason that representations are important is their convenience. All mathematicians know how to multiply matrices, and thus a group of matrices is easier to describe than some more abstractly defined group. But the importance of Lie group representations goes far beyond mere convenience.

For example, representations of A_1—the first group in Killing's table—explain the structure of the periodic table in chemistry. Surely every chemistry student has wondered why the rows of the periodic table contain 2, 8, 18, or 32 elements. Why did nature pick these numbers—each of which happens

Not only is E_8 larger than the other exceptional groups (which have dimensions 14, 52, 78, and 133 respectively), but it also *contains* all of the others.

All of the classical Lie groups can be expressed as a group of matrices in many different ways. Thus the problem arises of finding the "building blocks" or *irreducible representations*, from which the others can be constructed.

to be twice a square (1, 4, 9, 16)? The answer is that the number of elements in each row corresponds to the number of possible "shapes" of electron orbits in the atom's outer orbital. These shapes can be expressed by functions called spherical harmonics (see Figure 1). Each spherical harmonic is the graph in spherical coordinates of a polynomial, whose degree corresponds to the angular momentum of the electron. The space of all spherical harmonics of a given degree, or angular momentum, is an irreducible representation of A_1. The dimensions of these spaces combine to generate the seemingly magic numbers $2 (= 1 + 1)$, $8 (= 1 + 1 + 3 + 3)$, $18 (= 1 + 1 + 3 + 3 + 5 + 5)$, and so on. They would continue on to 50 and beyond, if only the corresponding atoms were stable.

Other Lie group representations have also played a prominent role in physics. The fundamental symmetry group in the theory of quarks is the compact version of A_2, better known as SU(3). Because A_1 accounts for one Nobel-caliber discovery (the periodic table) and A_2 produces another (the theory of quarks), it is understandable that some physicists have tried to jump to the end of the line, advocating "theories of everything" based on E_8. One version of string theory, which has for years been physicists' best hope to unify gravity with the other fundamental forces of nature, uses the compact version of E_8. In 2007, Garrett Lisi published an alternative to string theory based on a non-compact form of E_8. Lisi's theory attracted a great deal of media attention, in part due to his iconoclastic, surfer-dude personality. It is still too early to determine whether such attempts will be successful, but at least they have a distinguished pedigree.

All of the classical Lie groups can be expressed as a group of matrices in many different ways. Thus the problem arises of finding the "building blocks" or *irreducible representations*, from which the others can be constructed. In the 1920s, a German mathematician, Hermann Weyl, completely solved this problem, subject to one very important assumption: The Lie group had to be compact. This means, roughly speaking, that the group does not stretch out infinitely far in any direction. Without this assumption, everything breaks down.

Here is how Weyl found the irreducible representations. One of the key concepts in abstract algebra is the *canonical form* of a matrix. If A is a matrix, its canonical form is a matrix conjugate to A (in other words, it can be written as BAB^{-1}) that has a particularly nice structure. For instance, the canonical form may consist of nothing but 0's except along the main diagonal. Weyl showed that the same idea generalizes to *any* compact Lie group. The "diagonal" elements form a subgroup known as a *maximal torus* of the Lie group—a very appropriate name because it is topologically a torus (that is, an r-fold Cartesian product of circles). The dimension of this torus, r, is the rank of the Lie group, the mysterious subscript that appears in all the names of the simple Lie groups. Thus, for instance, A_1 has rank 1; G_2 has rank 2; and E_8 has rank 8. This means that somewhere in E_8, there is a subgroup T that looks just like an 8-dimensional torus.

Weyl realized that the maximal torus was the key to understanding the representations of a Lie group. Any representation of the whole group is automatically a representation of

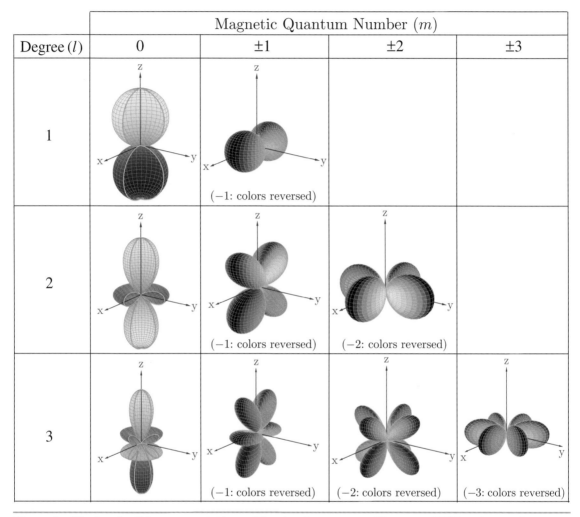

Degree (l)	Magnetic Quantum Number (m)			
	0	±1	±2	±3
1		(−1: colors reversed)		
2		(−1: colors reversed)	(−2: colors reversed)	
3		(−1: colors reversed)	(−2: colors reversed)	(−3: colors reversed)

Figure 1. *The irreducible representations of the simplest compact Lie group, SO(3) or A_1, are the spaces of spherical harmonics of degree n. Pictured here are the spherical harmonics of degrees 1, 2, and 3. Each spherical harmonic is a complex-valued function of the two spherical coordinates, θ and ϕ. In these graphs, the real part of the function provides the radial coordinate, while the imaginary part provides the color of the surface. One spherical harmonic of each degree is purely real (and thus plotted in gray), while the other $2n$ harmonics come in conjugate pairs. (Figures courtesy of Andrew D. Hwang.)*

the maximal torus; and representations of tori are very easy to understand. For the 1-dimensional torus (a circle), they are 1×1 matrix-valued functions of the form $[e^{2\pi ix}]$, $[e^{4\pi ix}]$, or in general $[e^{2\pi inx}]$ for any integer n (and where x represents the angular coordinate on the circle). Thus there is a direct correspondence between representations of a circle and integers, n, which can be thought of as points in a one-dimensional lattice.

Similarly, representations of an r-dimensional torus correspond to a lattice in r-dimensional space. The question then becomes: Which lattice points actually correspond to representations of the whole Lie group? And which of the representations are irreducible?

To answer this, remember that the Lie group has many more dimensions that lie outside of the maximal torus. For instance, G_2 has 14 dimensions, only 2 of which appear in the maximal torus. Similarly, E_8 has 248 dimensions, only 8 of which lie in the maximal torus. Each of the missing dimensions corresponds to a particular vector in the r-dimensional lattice, called a *root*. Figure 2 shows the 12 roots of G_2, which form a six-pointed star in the plane. The 240 roots of E_8 form the remarkable polyhedron that was described previously (**Symphony in 2, 3, and 5**), which generates the densest possible packing of spheres in eight-dimensional space.

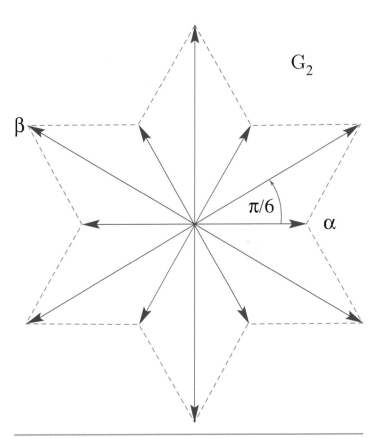

Figure 2a. *The root system of the simplest exceptional Lie group, G_2. The twelve roots from a six-pointed star, which is symmetric about the perpendicular to any of the root vectors.*

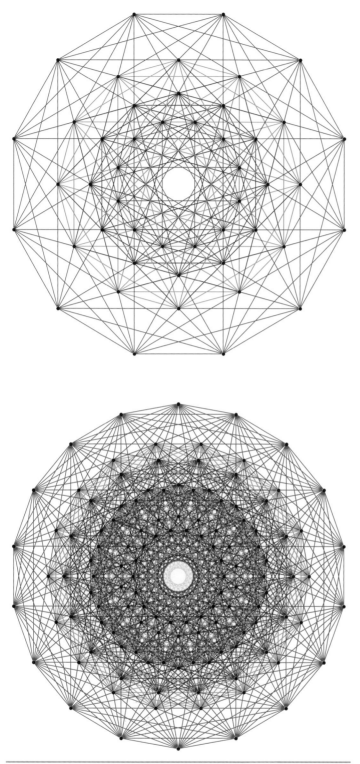

Figure 2b. *The root systems of E_6 and E_7. These lie in 6- and 7-dimensional space, respectively, and here have been projected into a plane that preserves as much as possible of their symmetry. (Compare* **Symphony in 2, 3, and 5,** *page 98.) (Figure courtesy of John Stembridge, University of Michigan.)*

As elegant as it was, Weyl's classification of irreducible representations worked only for compact groups. But many interesting Lie groups are not compact.

The roots always generate a discrete lattice in r-dimensional space, and the points in this lattice correspond to representations of the entire Lie group (not just the torus). The roots have rather special symmetries, which are inherited by the root lattice. In particular, the entire root lattice is symmetric about a mirror placed perpendicular to any root. Thus the whole root lattice can be thought of as a kaleidoscopic image of one "chamber" in a hall of mirrors. The E_8 lattice has 240 mirrors (one perpendicular to each root), which divide the space up into 696,729,600 identical chambers. The *irreducible* representations of E_8 correspond to the lattice points in *one* of those chambers.

As elegant as it was, Weyl's classification of irreducible representations worked only for compact groups. But many interesting Lie groups are not compact. One simple example is $SL_2(\mathbf{R})$, the set of real 2×2 matrices $\begin{bmatrix} a & b \\ c & d \end{bmatrix}$ such that the determinant $(ad - bc) = 1$. This group is non-compact because the numbers a, b, c, d can become arbitrarily large. Incidentally, this group contains a discrete subgroup called the modular group (see "A New Twist in Knot Theory" and "Error-Term Roulette and the Sato-Tate Conjecture"), which has always been of great interest to number theorists.

A second example, important to physicists, is $SO(2, 1)$. It is easiest to understand by analogy with $SO(3)$, the group of rotations of 3-space, which can be described as the set of linear transformations that preserve the squared length of any vector $\mathbf{v}$. If $\mathbf{v} = (x, y, z)$, then the squared length of $\mathbf{v}$, according to the Pythagorean theorem, is $|\mathbf{v}|^2 = x^2 + y^2 + z^2$.

In relativity theory, physicists make one slight alteration to the definition of length. Suppose that you have two dimensions of space (x, y), and one dimension of time (t). According to Albert Einstein's theory of special relativity, you can wrap all three dimensions together into a single spacetime, in which the squared length of any vector $\mathbf{v} = (x, y, t)$ is given by $|\mathbf{v}|^2 = x^2 + y^2 - t^2$. The only modification is the minus sign that distinguishes the time coordinate from the space coordinates. The *Lorentz group* $SO(2, 1)$, or the group of symmetries of this three-dimensional spacetime, is the set of linear transformations that preserve this modified length function. Just as $SO(3)$ can be thought of as rotating a unit sphere, $SO(2, 1)$ can be thought of as generating a flow on a unit hyperboloid (Figure 3). Because the hyperboloid is open-ended, $SO(2, 1)$ is non-compact.

Both $SL_2(\mathbf{R})$ and $SO(2, 1)$ look microscopically, or "infinitesimally," the same, and they are closely related to $SO(3)$. They are all three-dimensional, and if you take a very small neighborhood of the identity element, $\begin{bmatrix} 1 & 0 \\ 0 & 1 \end{bmatrix}$ or $\begin{bmatrix} 1 & 0 & 0 \\ 0 & 1 & 0 \\ 0 & 0 & 1 \end{bmatrix}$, respectively, the matrices in that neighborhood will multiply in essentially the same way. Thus the non-compact groups are considered alternate forms of the Lie group A_1.

In non-compact Lie groups, Weyl's construction of the maximal torus breaks down in two important ways. First, some of the circles that form the torus may open up into straight lines, making it non-compact. (In fact, from a topologist's point of view, it's not even a torus any more.) In addition, the maximal "torus" is no longer unique. Instead, the Lie group contains many different families of maximal tori. The tori within each family are conjugate to one another, but tori in separate families are not. Each of these families contributes a distinct set of irreducible representations. Thus, the Atlas of Lie Groups project had to identify and keep track of all of these different families.

Non-compact Lie groups are also different from compact ones in another respect: Their representations are usually not finite-dimensional. Also, they may not be unitary, which means that the group does not preserve any kind of length measurement. This has serious consequences both for mathematics and physics. Mathematically, the idea of irreducible subspaces as building blocks becomes much more subtle. Also, only unitary representations are considered by quantum physicists to be "physically meaningful" because they prevent negative probabilities from appearing.

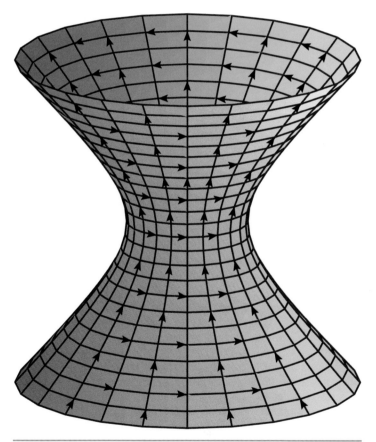

Figure 3. *(In the Lorentz metric, a "unit ball" is a hyperboloid. The symmetry operations of the hyperboloid have unbounded orbits (indicated by upward pointing arrows). For this reason, the Lie group of linear transformations preserving the Lorentz metric is not compact. (Figure courtesy of David Vogan.)*

To simplify their task, Lie group theorists often restrict their attention to *reductive* Lie groups. These are non-compact groups that locally look like the compact ones. We have seen two examples already: $SL_2(\mathbf{R})$ and $SO(2, 1)$, which "look like" A_1. As these examples show, there may be more than one non-compact group of the same type. The exceptional group E_8, for instance, comes in three different flavors: the classic, compact version; the "quaternionic" version, and the "split" version. All of them are 248-dimensional, have rank 8, and have the same root lattice. But on a large scale, they are put together differently.

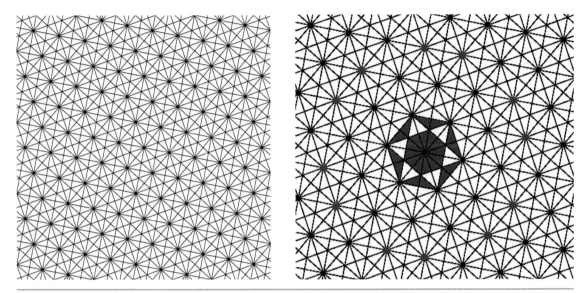

Figure 4. *(left) The root system of G_2 can be used to generate a lattice in the root space. For the non-compact version of G_2, irreducible representations correspond to points that are not on the lines in this lattice. (Figure courtesy of John Stembridge.) (right) For the non-compact version of G_2, unitary representations correspond to the points colored red in this figure. For E_8, the problem of finding a similarly concise description of the unitary representations remains unsolved. (Figure courtesy of Jeffrey Adams.)*

When Adams started the Atlas of Lie Groups project in 2002, his central question was this: Can we identify all of the irreducible, unitary representations of any non-compact, reductive Lie group? Usually, when mathematicians ask questions like this, they mean it in a very abstract, philosophical way. It's good enough to exhibit an algorithm that would work, in theory, on an arbitrarily large computer with an unlimited amount of time. But Adams meant it literally. Are these computations feasible, with currently existing computers, in a reasonable time?

A few special cases had been solved already. In 1947, Valentine Bargmann, a German physicist, had computed the representations of $SL_2(\mathbf{R})$. He showed that the natural parametrization of the family included a discrete lattice (as in the compact case) and a continuous family, and the continuous family could be indexed by complex numbers. Almost all of these are irreducible; however, only a few are unitary, namely the ones corresponding to purely imaginary numbers or to real numbers between -1 and 1.

The first non-compact exceptional group to have its unitary representations classified was the smallest one, the split form of G_2. Its representations can be viewed in the same root space as the compact form of G_2. In the compact form, the vectors on the root lattice correspond to irreducible representations and the other vectors in the root space do not correspond to anything. In the split form, every representation in the root space that does *not* lie on one of the lines in Figure 4 corresponds to an irreducible representation. This was proved by Birgit Speh of Cornell University in her 1977 doctoral thesis. In 1994, David Vogan of the Massachusetts Institute of Technology proved that only the representations that lie in the red shaded area (including the boundaries) are unitary.

In general, as mentioned above, irreducible representations will correspond to families of "straightened-up maximal tori." If d of the r circles (for a Lie group of rank r) have straightened up into lines, then the representations corresponding to that family will form a d-dimensional complex manifold. For example, if no circles straighten up, the representations form a 0-dimensional manifold, in other words a lattice of isolated points. (That is exactly what Weyl showed in the compact case.) If one circle straightens up, the representations form a 1-dimensional complex manifold. (That is what happened in the case of $\mathrm{SL}_2(\mathbf{R})$). In general, if the "torus" has d lines and $(r - d)$ circles as its factors, then the representations are described by a Cartesian product of a discrete lattice and a d-dimensional complex manifold.

In 1981, Alexander Beilinson and Joseph Bernstein (who were then in Moscow) proved that the families form a distinctive hierarchical structure (see Figure 5). Even though different families are not conjugate, adjacent families in the hierarchy are related to each other, and this fact turned out to be crucial for making the "map" of E_8.

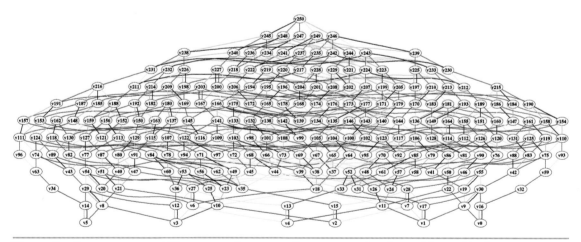

Figure 5. *For non-compact Lie groups, the space of representations has a hierarchical structure composed of pieces with different dimensions. The larger pieces "join" the smaller pieces together in the same way that a tent joins the pieces of the framework over which it is draped. In this figure, the different pieces are identified by numbers, and the lines connect higher-dimensional pieces to the lower-dimensional pieces they join. The figure illustrates the hierarchical structure for the non-compact group SO(5, 5); a similar figure for E_8 would require many more sheets of paper. (Figure courtesy of Scott Crofts.)*

> **... the final matrix still amounted to 60 gigabytes of data, which is more than 60 times the amount of data in the human genome. If you were to write the entire matrix out in normal-sized type, it would cover the island of Manhattan.**

In the mid-1980s, Vogan described a finite algorithm to compute the irreducible representations of a non-compact reductive group. His algorithm shows how to represent certain easily identified *standard representations* (one per family) as a linear combination of a less easily identified set of *irreducible representations* (also one per family). If you "reverse engineer" this algorithm, then you can define the (unknown) irreducible transformations in terms of the (known) standard ones. The transformation from standard to irreducible, or vice versa, is represented by an ordinary matrix of numbers. However, mathematicians before Vogan did not know how to calculate that matrix. Vogan succeeded by rewriting the entries of the matrix as *polynomials* instead of numbers (they became known as Kazhdan-Lusztig-Vogan or KLV polynomials). The hierarchical, tree-like structure of the families gave him a completely automatic procedure for computing the matrix entries, basically by ascending and descending the branches of the tree.

At that point, the abstract mathematical problem was solved. If you had a big enough computer, and if the universe lasted long enough, you could compute the irreducible representations of E_8 or any other reductive Lie group. But this was like telling Columbus that if he sailed west long enough, he would in theory get to India. How long was long enough? And what surprises would he encounter on the way? In many ways, the fun was only beginning.

When the Atlas team began work in 2002, it soon became apparent that the split form of E_8 was the main obstacle to completing Vogan's program. Compared to it, all of the other reductive Lie groups were mere warm-up exercises. It has 453,060 different families of maximal tori, and thus 453,060 families of representations. Its KLV polynomials form a $453,060 \times 453,060$ matrix, with more than 200 billion entries. Even though quite a few of these polynomials are equal to 0, the final matrix still amounted to 60 gigabytes of data, which is more than 60 times the amount of data in the human genome. If you were to write the entire matrix out in normal-sized type, it would cover the island of Manhattan.

But the nickname "a calculation the size of Manhattan," which was the title of a press release describing the work, actually underestimates the complexity of the computation! Any math student knows that it takes more paper to *figure out an answer* than simply to write it down. It was the act of performing the computation, not writing down the solution, that challenged the limits of modern-day computers.

To run Vogan's algorithm successfully, a computer had to retain the hierarchical graph in its RAM (easily accessed memory) all the time. The team tried storing the data on the hard disk, but that turned out to be way too slow: the computer would spend almost all of its time looking up data instead of performing calculations. At first, the team did not know how many gigabytes of RAM they would need. They knew how many KLV polynomials there were, but they did not know how large the polynomials might turn out to be. Early estimates suggested that they might need up to a terabyte of RAM, although in the final analysis they needed "only" 60 gigabytes.

The first few attempts to compute the character table were far from promising. Birne Binegar of Oklahoma State University burned out his computer's hard disk controller and disk drives, and that taught him the hard way that storing the data on disk was not the right way to go. "After this incident, E_8 became my Moby Dick," he says. Oklahoma State had just bought a supercomputer with 64 processors and 256 gigabytes of RAM. In theory that should have been enough, but when Binegar set all 64 processors to work on the E_8 calculation at the same time, a funny thing happened. "After about 40 minutes the system manager noticed that the air-conditioned server room was getting awfully hot," Binegar says. "In fact, it was getting upwards of 90 degrees and threatening the other critical servers in that room." The administrator turned the computer off and warned Binegar, "Don't ever do that again!"

Discouraged, Binegar and Adams started to think about buying a (single-processor) computer with 256 gigabytes of RAM, which would have cost about $150,000. But that turned out to be unnecessary, for two main reasons. The first, and most important, was the programming genius of Fokko du Cloux.

Du Cloux had already written sensationally fast programs to compute representations for another kind of group called Coxeter groups, and he was probably the only person in the world who knew both the mathematics and the computer programming well enough to make Vogan's algorithm practical. One problem was that when mathematicians prove theorems (such as the existence of this algorithm), they take short cuts. They know sometimes what "ought" to be true and don't fill in every last detail. As Vogan says, du Cloux had to "take it from reasonable mathematics to perfect." At the same time, he also had to streamline the code relentlessly, squeezing the most out of every last bit of RAM. He streamlined it enough to run on a 64-gigabyte computer at the University of Washington. Thus he saved the Atlas project (and the National Science Foundation) one hundred fifty thousand dollars!

Du Cloux's feat was especially remarkable because he was diagnosed with amyotrophic lateral sclerosis (a fatal, degenerative disease of the nervous system) in November 2005. He was completely paralyzed by the following spring. "He would lie on his back in Lyons (France), with a video projector pointing at the ceiling," says Adams. "I would type here in Baltimore, and he would see on the ceiling what I was typing. We would use Skype [an Internet telephone program] to talk." In spite of his condition, Adams adds, "All he wanted to talk about was mathematics. He never complained about this terrible thing that had happened to him."

Still, there was one final ingredient needed before the Atlas computation could work—a trick that is at least 1600 years old! Noam Elkies, a number theorist at Harvard University, asked whether they had thought of using the Chinese Remainder Theorem (see Sidebar). In essence, this trick parallelizes the computation of large numbers, by performing the computation modulo several smaller numbers. Vogan started to write a reply to Elkies explaining why it wouldn't work—and then he realized that it would!

Figure 5. *Fokko du Cloux played a unique role in the Atlas project, by virtue of his programming prowess and the inspiration he provided as he battled his fatal illness. (Photo courtesy of Bill Casselman.)*

A New Problem with an Ancient Solution

Calculating the irreducible representations of E_8 involved very sophisticated, contemporary mathematics. So it came as a big surprise when the very last step—which, in effect, divided the calculation into four more manageable pieces—drew upon a trick that was centuries old, the Chinese Remainder Theorem.

Suppose that you have a very large number of objects to count—say, 11,808,808 apples—and you want to break the problem up into more manageable pieces. One thing you can do is pack them into boxes that hold 251 apples apiece and see how many are left over (answer: 11). Then you could repeat it with boxes that hold 253 apples apiece. (Now there are 33 left over). Repeat again with boxes that hold 255 apples apiece. (Now there are 13 left over.) Finally, repackage them into boxes that hold 256 apples apiece. (There are 40 left over.)

Now comes the magic: Only one number less than 4 billion (to be precise, less than $251 \times 253 \times 255 \times 256$) has the specified remainders when divided by those four numbers. There is a simple number-theoretic algorithm to compute it. Thus, provided you know at the outset that you have fewer than 4 billion apples, the Chinese Remainder Theorem tells you that you have exactly 11,808,808. So you can find the number of apples without ever counting past 256!

A version of this trick appeared in a Chinese math book in the fourth century AD. The specific numbers used there were 3, 5, and 7, but the Chinese Remainder Theorem will work with any set of numbers that have no common divisors. The numbers 251, 253, 255, and 256 were convenient for the mathematicians in the Atlas Project because the computers could use eight-bit arithmetic (that is, integers less than $2^8 = 256$) to figure out the remainders.

Of course, the Atlas team was not counting apples but calculating coefficients of the KLV polynomials. They estimated beforehand that none of the coefficients would be larger than 4 billion. In the end, it turned out that the largest coefficient that appeared in any of the KLV polynomials was 11,808,808.

Elkies' suggestion reduced a "calculation the size of Manhattan" to four calculations that were each the size of Cambridge, Massachusetts. That was simplification enough. Despite several computer crashes and an inconvenient interruption due to Christmas, the four calculations were completed in December 2006 and January 2007, and the complete matrix of KLV polynomials was written to a computer disk on January 8. Unfortunately, du Cloux did not live to see the result of the computation—he had died on November 10.

Without du Cloux, the Atlas project has some large shoes to fill. Marc van Leeuwen, a mathematician at the University of Poitiers who wrote the software to implement the Chinese Remainder Theorem, has taken on the challenge of understanding du Cloux's code and adding to it. For example, he is working on software to compute the restriction of infinite-dimensional representations of a Lie group to its maximal compact subgroups. "That ability will allow us to investigate unitary representations in interesting new ways," Vogan says.

In fact, one of the main goals of the Atlas project—to compute the unitary representations—remains incomplete. It is still not clear when or whether it will be finished. Nevertheless, the data from the character table have already enabled members of the Atlas team to answer some previously untouchable questions. It does more than identify the irreducible representations; it also contains geometric information about them. For example, even though the representations are infinite-dimensional, it is possible to assign to them a degree of infinite-dimensionality, called the Gelfand-Kirillov dimension, which for E_8 is a number between 1 and 120. The Gelfand-Kirillov dimension can now be computed for every representation of E_8.

Perhaps more importantly, the patterns revealed in the character table may inspire new questions. For example, it turned out unexpectedly that all of the KLV polynomials for E_8 have constant terms equal to 0, 1, 2, 4, or 8—a pattern that currently defies explanation. "There are lots and lots of little surprises," Vogan says. While this particular observation may be of no great significance, the atlas will undoubtedly make chance discoveries possible that were not before. And if history is any guide, the most wonderful things about a new world are the things that you never expected to find.

. . . the atlas will undoubtedly make chance discoveries possible that were not before. And if history is any guide, the most wonderful things about a new world are the things that you never expected to find.

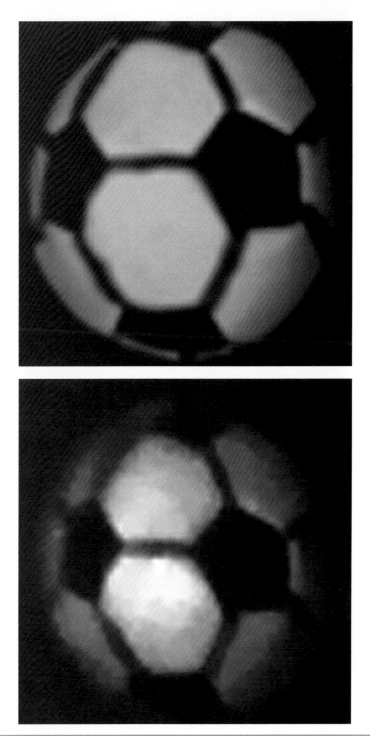

One is Enough. *A photograph taken by the "single-pixel camera" built by Richard Baraniuk and Kevin Kelly of Rice University. (a) A photograph of a soccer ball, taken by a conventional digital camera at* 64×64 *resolution. (b) The same soccer ball, photographed by a single-pixel camera. The image is derived mathematically from 1600 separate, randomly selected measurements, using a method called compressed sensing. (Photos courtesy of R. G. Baraniuk, Compressive Sensing [Lecture Notes],* Signal Processing Magazine, *July 2007.* © *2007 IEEE.)*

Compressed Sensing Makes Every Pixel Count

TRASH AND COMPUTER FILES HAVE ONE THING in common: compact is beautiful. But if you've ever shopped for a digital camera, you might have noticed that camera manufacturers haven't gotten the message. A few years ago, electronic stores were full of 1- or 2-megapixel cameras. Then along came cameras with 3-megapixel chips, 10 megapixels, and even 60 megapixels.

Unfortunately, these multi-megapixel cameras create enormous computer files. So the first thing most people do, if they plan to send a photo by e-mail or post it on the Web, is to compact it to a more manageable size. Usually it is impossible to discern the difference between the compressed photo and the original with the naked eye (see Figure 1, next page). Thus, a strange dynamic has evolved, in which camera engineers cram more and more data onto a chip, while software engineers design cleverer and cleverer ways to get rid of it.

In 2004, mathematicians discovered a way to bring this "arms race" to a halt. Why make 10 million measurements, they asked, when you might need only 10 thousand to adequately describe your image? Wouldn't it be better if you could just acquire the 10 thousand most relevant pieces of information at the outset? Thanks to Emmanuel Candes of Caltech, Terence Tao of the University of California at Los Angeles, Justin Romberg of Georgia Tech, and David Donoho of Stanford University, a powerful mathematical technique can reduce the data a thousandfold *before* it is acquired. Their technique, called *compressed sensing*, has become a new buzzword in engineering, but its mathematical roots are decades old.

As a proof of concept, Richard Baraniuk and Kevin Kelly of Rice University even developed a *single-pixel* camera. However, don't expect it to show up next to the 10-megapixel cameras at your local Wal-Mart because megapixel camera chips have a built-in economic advantage. "The fact that we can so cheaply build them is due to a very fortunate coincidence, that the wavelengths of light that our eyes respond to are the same ones that silicon responds to," says Baraniuk. "This has allowed camera makers to jump on the Moore's Law bandwagon"—in other words, to double the number of pixels every couple of years.

Thus, the true market for compressed sensing lies in non-visible wavelengths. Sensors in these wavelengths are not so cheap to build, and they have many applications. For example, cell phones detect encoded signals from a broad spectrum of radio frequencies. Detectors of terahertz radiation[1] could be used to spot contraband or concealed weapons under clothing.

Emmanuel Candes. *(Photo courtesy of Emmanuel Candes.)*

---

[1]This is a part of the electromagnetic spectrum that could either be described as ultra-ultra high frequency radio or infra-infrared light, depending on your point of view.

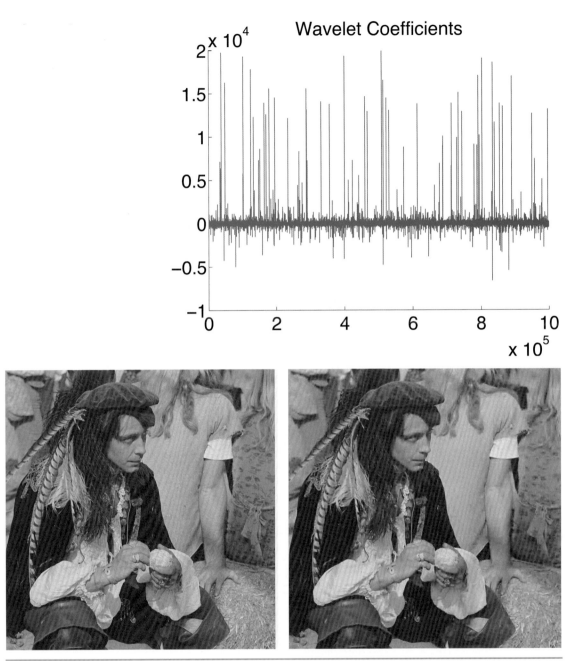

Figure 1. *Normal scenes from everyday life are compressible with respect to a basis of wavelets. (left) A test image. (top) One standard compression procedure is to represent the image as a sum of wavelets. Here, the coefficients of the wavelets are plotted, with large coefficients identifying wavelets that make a significant contribution to the image (such as identifying an edge or a texture). (right) When the wavelets with small coefficients are discarded and the image is reconstructed from only the remaining wavelets, it is nearly indistinguishable from the original. (Photos and figure courtesy of Emmanuel Candes.)*

Even conventional infrared light is expensive to image. "When you move outside the range where silicon is sensitive, your $100 camera becomes a $100,000 camera," says Baraniuk. In some applications, such as spacecraft, there may not be enough room for a lot of sensors. For applications like these, it makes sense to think seriously about how to make every pixel count.

The Old Conventional Wisdom

The story of compressed sensing begins with Claude Shannon, the pioneer of information theory. In 1949, Shannon proved that a time-varying signal with no frequencies higher than N hertz can be perfectly reconstructed by sampling the signal at regular intervals of $1/2N$ seconds. But it is the converse theorem that became gospel to generations of signal processors: a signal with frequencies higher than N hertz *cannot* be reconstructed uniquely; there is always a possibility of aliasing (two different signals that have the same samples).

Terence Tao. *(Photo courtesy of Reed Hutchinson/UCLA.)*

In the digital imaging world, a "signal" is an image, and a "sample" of the image is typically a pixel, in other words a measurement of light intensity (perhaps coupled with color information) at a particular point. Shannon's theorem (also called the Shannon-Nyquist sampling theorem) then says that the resolution of an image is proportional to the number of measurements. If you want to double the resolution, you'd better double the number of pixels. This is exactly the world as seen by digital-camera salesmen.

Candes, Tao, Romberg, and Donoho have turned that world upside down. In the compressed-sensing view of the world, the achievable resolution is controlled primarily by the *information content of the image.* An image with low information content can be reconstructed perfectly from a small number of measurements. Once you have made the requisite number of measurements, it doesn't help you to add more. If such images were rare or unusual, this news might not be very exciting. But in fact, *virtually all real-world images have low information content* (as shown in Figure 1).

This point may seem extremely counterintuitive because the mathematical meaning of "information" is nearly the opposite of the common-sense meaning. An example of an image with high information content is a picture of random static on a TV screen. Most laymen would probably consider such a signal to contain no information at all! But to a mathematician, it has high information content precisely because it has no pattern; in order to describe the image or distinguish between two such images, you literally have to specify every pixel. By contrast, any real-world scene has low information content because it is possible to convey the content of the image with a small number of descriptors. A few lines are sufficient to convey the idea of a face, and a skilled artist can create a recognizable likeness of any face with a relatively small number of brush strokes.[2]

---

[2]The modern-day version of the "skilled artist" is an image compression algorithm, such as the JPEG-2000 standard, which reconstructs a copy of the original image from a small number of components called *wavelets*. (See "Parlez-vous Wavelets?" in *What's Happening in the Mathematical Sciences*, Volume 2.)

Justin Romberg. *(Photo courtesy of Justin Romberg.)*

The idea of compressed sensing is to use the low information content of most real-life images to circumvent the Shannon-Nyquist sampling theorem. If you have no information at all about the signal or image you are trying to reconstruct, then Shannon's theorem correctly limits the resolution that you can achieve. But if you know that the image is sparse or compressible, then Shannon's limits do not apply.

Long before "compressed sensing" became a buzzword, there had been hints of this fact. In the late 1970s, seismic engineers started to discover that "the so-called fundamental limits weren't fundamental," says Donoho. Seismologists gather information about underground rock formations by bouncing seismic waves off the discontinuities between strata. (Any abrupt change in the rock's state or composition, such as a layer of oil-bearing rock, will reflect a vibrational wave back to the surface.) In theory the reflected waves did not contain enough information to reconstruct the rock layers uniquely. Nevertheless, seismologists were able to acquire better images than they had a right to expect. The ability to "see underground" made oil prospecting into less of a hit-or-miss proposition. The seismologists explained their good fortune with the "sparse spike train hypothesis," Donoho says. The hypothesis is that underground rock structures are fairly simple. At most depths, the rock is homogeneous, and so an incoming seismic wave sees nothing at all. Intermittently, the seismic waves encounter a discontinuity in the rock, and they return a sharp spike to the sender. Thus, the signal is a sparse sequence of spikes with long gaps between them.

In this circumstance, it is possible to beat the constraints of Shannon's theorem. It may be easier to think of the dual situation: a sparse *wave train* that is the superposition of just a few sinusoidal waves, whose frequency does not exceed N hertz. If there are K frequency spikes in a signal with maximal frequency N, Shannon's theorem would tell you to collect N equally spaced samples. But the sparse wave train hypothesis lets you get by with only $3K$ samples, or even sometimes just $2K$. The trick is to sample at random intervals, not at regular intervals (see Figures 2 and 3). If $K \ll N$ (which is the meaning of a "sparse" signal), then random sampling is much more efficient.

In other fields, such as magnetic resonance imaging, researchers also found that they could "undersample" the data and still get good results. At scientific meetings, Donoho says, they always encountered skepticism because they were trying to do something that was supposed to be impossible. In retrospect, he says that they needed a sort of mathematical "certificate," a stamp of approval that would guarantee when random sampling works.

The New Certificate

Emmanuel Candes, a former student of Donoho, faced the same skepticism in 2004, while working with a team of radiologists on magnetic resonance imaging. In trial runs with a "phantom image" (in other words, not a real patient), he was able to reconstruct the image perfectly from undersampled data. "There was no discrepancy at all between the original and the reconstruction," Candes says. "I actually got into a bit of trouble, because they thought I was fudging."

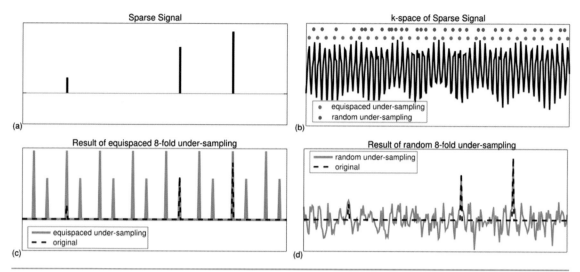

Figure 2. *Reconstructing a sparse wave train. (a) The frequency spectrum of a 3-sparse signal. (b) The signal itself, with two sampling strategies: regular sampling (red dots) and random sampling (blue dots). (c) When the spectrum is reconstructed from the regular samples, severe "aliasing" results because the number of samples is 8 times less than the Shannon-Nyquist limit. It is impossible to tell which frequencies are genuine and which are impostors. (d) With random samples, the two highest spikes can easily be picked out from the background. (Figure courtesy of M. Lustig, D. Donoho, J.Santos and J. Pauly,* Compressed Sensing MRI, Signal Processing Magazine, *March 2008. © 2008 IEEE.)*

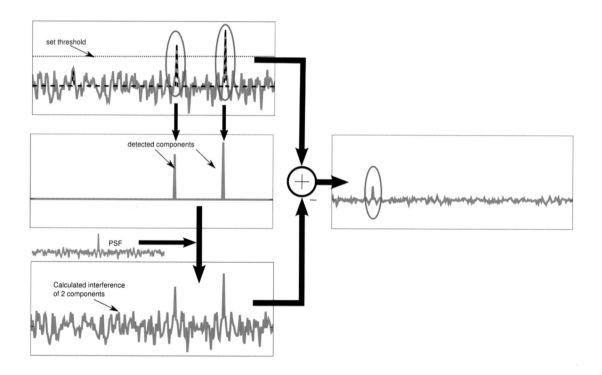

Figure 3. *In the situation of Figure 2, the third frequency spike can be recovered by an iterative thresholding procedure. If the signal was known to be 3-sparse to begin with, then the signal can be reconstructed perfectly, in spite of the 8-fold undersampling. In short, sparsity plus random sampling enables perfect (or near-perfect) reconstruction. (Figure courtesy of M. Lustig, D. Donoho, J.Santos and J. Pauly,* Compressed Sensing MRI, Signal Processing Magazine, *March 2008. © 2008 IEEE.)*

At this point, Candes did a fortuitous thing: he talked with Terry Tao, a 2006 Fields medalist. The two mathematicians happened to have children at the same pre-school. While they were dropping them off one day, Candes told Tao about the too-good-to-be-true reconstructions. "I had begun looking for an explanation and made some headway, but I was stuck at a particular point," Candes says.

"Terry reacted like a mathematician," Candes continues. "He said, 'I'm going to find a counterexample, showing that what you have in mind cannot be true.'" But a strange thing happened. None of the counterexamples seemed to work, and Tao started listening more closely to Candes' reasoning. "After a while, he looked at me and said, 'Maybe you're right,'" Candes says. With the speed for which Tao is legendary, within a few days he had helped Candes overcome his obstacle and the two of them began to sketch out the first truly general theory of compressed sensing.

In the Candes-Romberg-Tao framework, a signal or an image is represented as a vector $\mathbf{x}$, a string of N real numbers. This vector is assumed to be K-sparse, which means that in some prescribed basis it is known to have at most K nonzero coefficients. (K is assumed to be much less than N.) For example, if the basis elements are standard coordinate vectors in $\mathbf{R}^N$, then $\mathbf{x}$ literally consists of mostly zeroes. This is exactly the situation of the sparse spike train hypothesis.

However, compressed sensing does not require a particular basis. Photographs, for example, are not at all sparse with respect to the standard basis; they have many nonzero coefficients (i.e., non-black pixels). JPEG compression has proven that photographs are almost always approximately sparse with respect to a different basis—the basis of wavelets. If Ψ represents the $N \times N$ matrix of basis vectors, then a K-sparse signal with respect to that basis is one that can be written in the form $\Psi\mathbf{x}$, where $\mathbf{x}$ has at most K nonzero coefficients.

A sample $\mathbf{y}$ of the signal $\mathbf{x}$, in the Candes-Romberg-Tao framework, is a linear function of $\mathbf{x}$: that is, $\mathbf{y} = \Phi\mathbf{x}$. The number of measurements in the sample is assumed to be smaller than the signal, so Φ is an $M \times N$ matrix with $M << N$. By elementary linear algebra, there are infinitely many other vectors $\mathbf{x}^*$ such that $\Phi\mathbf{x}^* = \mathbf{y}$. However, provided that $M \geq 2K$, it will normally be the case that none of the other solutions to the equation $\Phi\mathbf{x}^* = \mathbf{y}$ are sparse. Thus, if $\mathbf{x}$ is known in advance to be sparse, it can in theory be reconstructed exactly from M measurements.

Knowing that a unique solution exists is not the same thing as being able to find it. The problem is that there is no way to know in advance which K coordinates of $\mathbf{x}$ are nonzero. The naive approach is to try all the possibilities until you hit on the right one, but this turns out to be a hopelessly slow algorithm. However, Candes and Tao found a shortcut that not only runs faster on a computer, but also explains why random sampling works so much better than regular sampling.

If your image consists of a few sparse dots or a few sharp lines, the *worst* way to sample it is by capturing individual pixels (the way a regular camera works!). The *best* way to sample the image is to compare it with widely spread-out noise functions. One could draw an analogy with the game of "20 questions." If you have to find a number between 1 and N, the worst way to proceed is to guess individual numbers (the analog of measuring individual pixels). On average, it will take you $N/2$

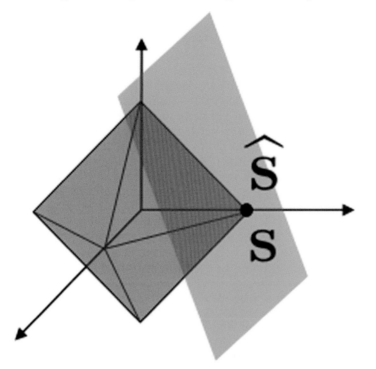

Figure 4. *A random measurement of a sparse signal, S, generates a subspace of possible signals (green) that could have produced that measurement. Within that green subspace, the vector of smallest l_1-norm ($\hat{S}$) is usually equal to S. (Figure courtesy of R. G. Baraniuk, Compressive Sensing [Lecture Notes], Signal Processing Magazine, July 2007. © 2007 IEEE.)*

guesses. By contrast, if you ask questions like, "Is the number less than $N/2$?" and then "Is the number less than $N/4$?" and so on, you can find the concealed number with at most $\log_2 N$ questions. If N is a large number, this is an enormous speed-up.

Notice that the "20 questions" strategy is adaptive: you are allowed to adapt your questions in light of the previous answers. To be practically relevant, Candes and Tao needed to make the measurement process nonadaptive, yet with the same guaranteed performance as the adaptive strategy just described. In other words, they needed to find out ahead of time what would be the most informative questions about the signal **x**. That this can be done effectively is one of the great surprises of the new theory. The idea of their approach is called l_1-minimization.

The l_0-norm of a vector is simply the number of nonzero entries in the vector, which can be somewhat informally written as follows:

$$\|(x_1, x_2, ..., x_N)\|_0 = \sum |x_i|^0.$$

(This formula uses the convention that $0^0 = 0$.) The l_0-norm is obtained by replacing the 0's in this equation by 1's:

$$\|(x_1, x_2, ..., x_N)\|_1 = \sum |x_i|^1.$$

In this language, the signal $\mathbf{x}$ is the unique solution to $\Phi \mathbf{x}^* = \mathbf{y}$ with the smallest l_0-norm. But in many cases, Candes and Tao proved, it is *also* the unique solution with the smallest l_1-norm. This was a critical insight because l_1-minimization is a linear programming problem, which can be solved by known, efficient computer algorithms. (See "Smooth(ed) Moves," *What's Happening in the Mathematical Sciences*, Volume 6.)

Figure 4 illustrates why the l_1-minimizer is often the same as the l_0-minimizer. In 3-dimensional space, the set of unit vectors in the l_1-norm is an octahedron. Think of the sparse vector $\mathbf{x}$ as lying on a coordinate axis (because it has lots of zero coordinates). Therefore it is at one of the vertices of the octahedron. The set of vectors $\mathbf{x}^*$ such that $\Phi \mathbf{x}^* = \mathbf{y}$ is a plane passing through the point $\mathbf{x}$. Most planes that pass through $\mathbf{x}$ intersect the octahedron *only* at the point $\mathbf{x}$; in other words, $\mathbf{x}$ is the unique point on the plane with the minimum l_1-norm. So if you simply pick the measurement Φ "at random," you have a very good chance of reconstructing $\mathbf{x}$ uniquely.

Unfortunately, picking Φ at random won't always work. You might get unlucky and choose a plane through $\mathbf{x}$ that passes through the interior of the octahedron. If so, the l_1-minimizer will not be the same as the l_0-minimizer. The algorithm will produce an erroneous signal, $\mathbf{x}^*$. But the three-dimensional picture in Figure 4 (page 121) is somewhat misleading because the image vectors typically lie in a space with thousands or millions of dimensions. The analog of the octahedron in million-dimensional space is called the cross polytope; and in million-dimensional space the cross polytope is very, very, very pointy. A random plane that passes through a vertex is *virtually certain* to miss the interior of the cross polytope. Thanks to this "miracle of high-dimensional geometry," as Candes calls it, the l_1-minimizer will almost always be the correct signal, $\mathbf{x}$.

In summary, this is what the theory of compressed sensing says:

- For many $M \times N$ matrices Φ, the unique K-sparse solution, $\mathbf{x}$, to the equation $\Phi \mathbf{x}^* = \mathbf{y}$, can be recovered *exactly*.
- N must be much larger than K. However, M (the number of measurements) need only be a little larger than K. Specifically, M must be roughly $K \log(N/K)$. Notice that the dependence on N is logarithmic, so the "20 questions" speed-up has been achieved.
- The K-sparse solution is found by l_1-minimization, which can be proved to be equivalent to l_0-minimization under certain assumptions on the measurement matrix, Φ.
- Random matrices Φ almost always satisfy those assumptions.

The whole story remains essentially unchanged if the signal is sparse with respect to a basis Ψ that is not the standard basis of coordinate vectors (e.g., the wavelet basis). The only

> A random plane that passes through a vertex is *virtually certain* to miss the interior of the cross polytope. Thanks to this "miracle of high-dimensional geometry," as Candes calls it, the l_1-minimizer will almost always be the correct signal, $\mathbf{x}$.

modification required is that the constraint, $\Phi x^* = y$, is replaced by the constraint $\Phi \Psi x^* = y$. In this context, the randomness of the measurement matrix Φ serves a double purpose. First, it provides the easiest set of circumstances under which l_1-minimization is provably equivalent to l_0-minimization. Secondly, and independently, it ensures that the set of measurement vectors (the rows of Φ) are as dissimilar to the image basis (the columns of Ψ) as possible. If the image basis consists of spikes, the measurement basis should consist of spread-out random noise. If the image basis consists of wavelets, then the measurement basis should consist of a complementary type of signal called "noiselets."

"Our paper showed something really unexpected," says Candes. "It showed that using randomness as a sensing mechanism is extremely powerful. That's claim number one. Claim number two is that it is amenable to rigorous analysis.

"What mathematicians liked [about the paper] was the way it merged analysis and probability theory. A lot of people in my field, analysis, did not think about probability theory as being useful or worthy of attention. At the very intellectual level, it changed the mindset of those people and caused them to engage this field."

Recent Developments

Tao and Candes' preprint appeared in 2004, as did a paper by Donoho announcing similar results. By the time that Tao and Candes' paper actually appeared in print, in 2006, it had been cited more than 100 times. Since then, there have been many advances, both from the theoretical and the practical side.

One question left unanswered by the original paper was how well compressed sensing would hold up if the measurements contained some random error (an inevitable problem of real-world devices), or if the images themselves were not exactly sparse. In photography, for instance, the assumption of a sparse signal is not literally true. It is more realistic to assume the signal is *compressible*, which means that the vast majority of the information in the signal is contained in a few coefficients. The remaining coefficients are not literally zero, but they are small. Under these circumstances, even the l_0-minimizer does not match the signal exactly, so there is no hope for the l_1-minimizer to be exactly correct.

In 2005, Candes, Romberg, and Tao showed that even with noisy measurements and compressible (but not sparse) signals, compressed sensing works well. The error in the reconstructed signal will not be much larger than the error in the measurements, and the error due to using the l_1-minimizer will not be much greater than the penalty already incurred by the l_0-minimizer. That is, the l_1-minimizer accurately recovers the most important pieces of information, the largest components of the signal. Figure 5 (see next page) shows an example of the performance of compressed sensing on a simulated image with added noise.

Mathematicians have also been working on new algorithms that run even faster than the standard linear programming techniques that solve the l_1-minimization problem. Instead of finding the largest K coefficients of x all at once, they find them iteratively: first the largest nonzero coefficient, then the

Richard Baraniuk. *(Photo courtesy of Richard Baraniuk.)*

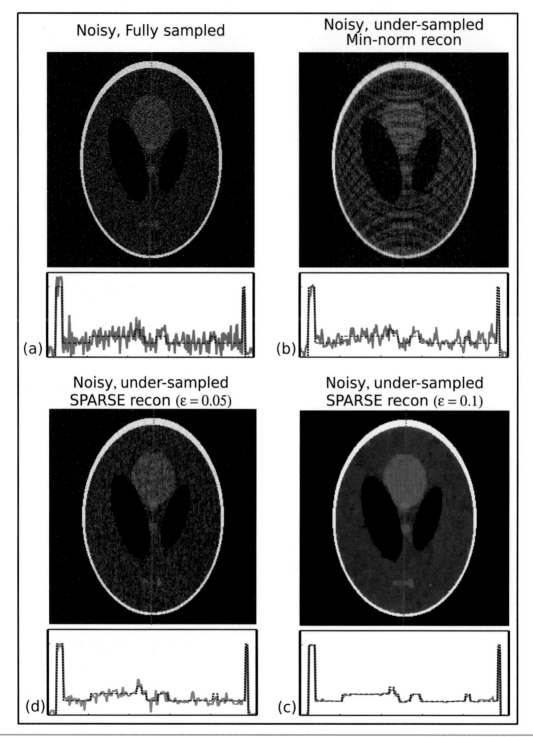

Figure 5. *Compressed sensing with noisy data. (a) An image with added noise. (b) The image, under-sampled and reconstructed using the Shannon-Nyquist approach. As in Figure 2, artifacts appear in the reconstructed image. (d) The same image, undersampled randomly and reconstructed with a "too optimistic" noise model. Although there are no artifacts, some of the noise has been misinterpreted as real variation. (c) The same image, reconstructed from a random sample with a more tolerant noise model. The noise is suppressed and there are no artifacts. (Figure courtesy of Michael Lustig.)*

second-largest one, and so on. The first such algorithm, called Orthogonal Matching Pursuit (OMP), did not offer the same guarantees of accuracy that l_1-minimization did. However, there is now a variety of colorfully named variations, such as Regularized OMP (ROMP) and Stagewise OMP (StOMP), which successfully combine the accuracy of l_1-minimization with the speed of OMP. These algorithms have the advantage of being somewhat more intuitive than the "high-dimensional miracle" of l_1-minimization; Figure 3 shows an example.

Meanwhile, researchers in several different fields are exploring practical applications of compressed sensing. Baraniuk and Kelly's single-pixel camera, built in 2006, uses an array of bacteria-sized mirrors to acquire a random sample of the incoming light. (See Figure 6.) Each mirror can be tilted in one of two ways, either to reflect the light toward the single sensor or away from it. Thus the light that the sensor receives is a weighted average of many different pixels, all combined into one pixel. By taking $K \log(N/K)$ snapshots, with a different random selection of pixels each time, the single-pixel camera was able to acquire a recognizable picture with a resolution comparable to N pixels. (See figure **"One Is Enough,"** page 114.)

Baraniuk and Kelly's team is now working on "hyperspectral cameras," which would reconstruct a complete spectrum at each point of the image. "A conventional digital image has red, blue and green pixels," Baraniuk says. "It's great for making a picture that fools the human eye, but it doesn't capture the essence of the wavelengths given off by different materials. What you'd really like would be a spectrum of thousands of colors instead of just three. This would allow you to tell the difference between green paint on a car and a green leaf on a bush." But with thousands of colors at each of millions of pixels, data compression becomes a serious issue.

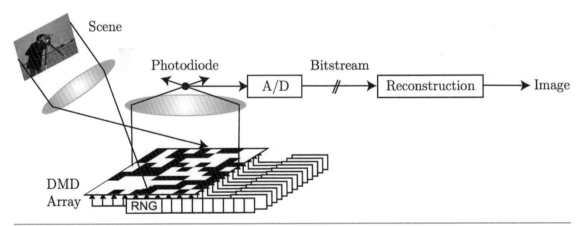

Figure 6. *A schematic diagram of the "one-pixel camera." The "DMD" is the grid of micro-mirrors that reflect some parts of the incoming light beam toward the sensor, which is a single photodiode. Other parts of the image (the black squares) are diverted away. Each measurement made by the photodiode is a random combination of many pixels. In* **"One is Enough"** *(p.114), 1600 random measurements suffice to create an image comparable to a 4096-pixel camera. (Figure courtesy of Richard Baraniuk.)*

With so many questions and so many choices, it is impossible at present to say what the most successful application of compressed sensing will be. However, one thing is clear: Engineers are finally thinking outside the box of Shannon's theorem.

Baraniuk and his former student Michael Wakin, now at the University of Michigan, have also worked on a problem of object detection. For many applications, producing an actual photograph may not be as important as recognizing quickly what is there. For example, a security system may have to identify a face or a vehicle. For example, Baraniuk says, you could teach it to recognize the difference between a Corolla and a Porsche. The computer will have images of Corollas and Porsches stored in it, but the vehicle in front of the camera may be rotated in a way that does not precisely match the photos. In this application, the image vector has a different kind of sparse structure. Instead of lying on a coordinate K-plane, the vector will lie on a curved K-dimensional manifold in N-dimensional space. (In this case, K would be equal to 3.) In this context, Wakin showed that on the order of $K \log N$ measurements still suffice to make the call.

Some applications of compressed sensing may lie completely outside the realm of imaging. One such example is "analog to digital conversion," a fundamental aspect of wireless communications. For example, the CDMA cell phone standard takes a voice message, which contains sound frequencies up to 4096 hertz, and spreads it out over a radio spectrum that spans hundreds of thousands of hertz. The signal is sparse because it still contains only the information that was squeezed inside those 4096 hertz. So a detector that performs compressed sensing should be able to recover the signal more rapidly than a detector based on Shannon's theorem.

In digital photography, Moore's law lets you pack twice as many detectors on a chip every two years. But in the world of analog to digital conversion, Baraniuk says, "the equivalent figure of merit doubles every 6 to 8 years." So instead of waiting decades for a hardware solution, it really makes sense to solve the problem with software based on compressed sensing.

Finally, compressed sensing may find some medical applications—which would be only natural because the theory was directly inspired by a problem in magnetic resonance imaging. MRI scanners have traditionally been limited to imaging static structures over a short period of time, and the patient has been instructed to hold his or her breath. But now, by treating the image as a sparse signal in space and time, MRI scanners have begun to overcome these limitations and produce images, for example, of a beating heart. Figure 7 shows how a sparse reconstruction algorithm can provide a sharp image of the arteries in a patient's leg even with as many as 20 times less data than a conventional angiogram.

One hurdle that compressed sensing may have to overcome is how to develop practical "incoherent sensors." A single measurement, in compressed sensing, is an inner product of the incoming compressible signal with a random, noisy test signal. Baraniuk's single-pixel camera accomplishes the inner product by using mirrors to deflect certain parts of the light beam toward the sensor, while deflecting other parts away. In real applications, if the hardware that performs the incoherent measurements is more expensive than the array of sensors that it is designed to replace, then the economic case for compressed sensing will disappear.

With so many questions and so many choices, it is impossible at present to say what the most successful application of compressed sensing will be. However, one thing is clear: Engineers are finally thinking outside the box of Shannon's theorem.

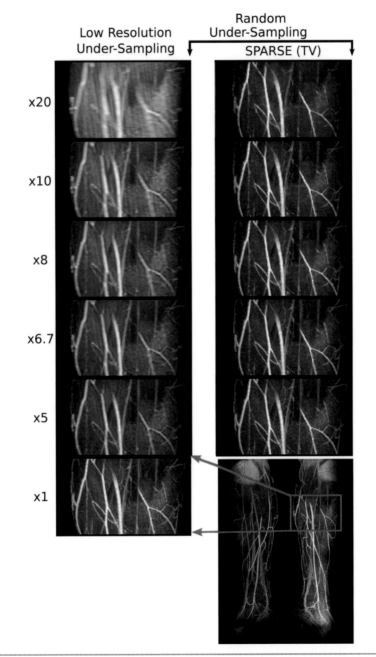

Figure 7. *An angiogram. From bottom to top, the angiogram is progressively undersampled by larger and larger factors. With a Shannon-Nyquist sampling strategy, the image degrades as the degree of undersampling increases. With compressed sensing, the image remains very crisp even at 20-fold undersampling. The approach used here and in Figure 5 is not l_1-minimization but l_1-minimization of the spatial gradient. (Figure courtesy of Michael Lustig.)*